NEO-BABYLONIAN ASTRONOMICAL TEXT.

86378,

OBVERSE, COLUMN I.

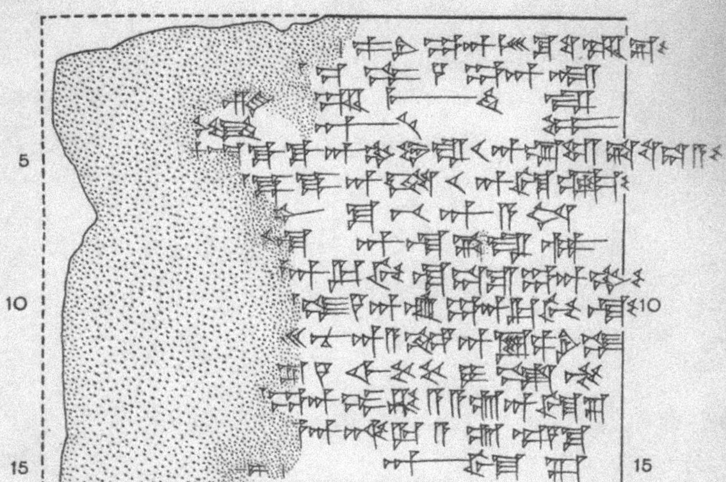

Writing Science before
the Greeks

# Culture and History of the Ancient Near East



VOLUME 48

# Writing Science before the Greeks

## A Naturalistic Analysis of the Babylonian Astronomical Treatise MUL.APIN

*By*

Rita Watson and Wayne Horowitz

TUTA SUB AEGIDE PALLAS · 1683 ·

BRILL

LEIDEN • BOSTON
2011

This book is printed on acid-free paper.

Library of Congress Cataloging-in-Publication Data

Watson, Rita
Writing science before the Greeks : a naturalistic analysis of the Babylonian astronomical treatise MUL.APIN / by Rita Watson and Wayne Horowitz.
p. cm. — (Culture and history of the ancient Near East, ISSN 1566-2055 ; v. 48)
Includes bibliographical references and index.
ISBN 978-90-04-20230-6 (hardback : alk. paper) 1. Astronomy, Assyro-Babylonian. 2. Akkadian language—Texts. I. Horowitz, Wayne, 1957– II. Title. III. Series.

QB19.W38 2011
520.935—dc22

2010051431

ISSN 1566-2055
ISBN 978 90 04 20230 6

In tribute to Herman Hunger and David R. Olson for their lifelong achievements in our respective fields; and in memory of our friend John Britton.

CONTENTS

## LIST OF ILLUSTRATIONS

## ACKNOWLEDGMENTS

The authors would like to acknowledge the editors of *Archiv für Orientforschung* and Herman Hunger for their permission to quote freely from the translated version of MUL.APIN in Hunger and Pingree 1989. We also acknowledge the Trustees of the British Museum for permission to study materials in the Museum's collection, and to reproduce copies of BM 86378. We would also like to express our thanks to the Hebrew University for support during the preparation of the manuscript, and to the numerous friends and colleagues who contributed their thoughts to our own during the incubation stage of the writing of this book; in particular, to David R. Olson for writing the foreword. Rita Watson also acknowledges support from The Abraham Schiffman Chair during the preparation of the manuscript, and Wayne Horowitz acknowledges The Israel Science Foundation for a research fellowship on the topic of Babylonian scholarship. We would also like to express our gratitude to our editors at Brill, Jennifer Pavelko and Katelyn Chin, for guiding the volume into print, and to Michael J. Mozina and Gene McGarry for their invaluable contribution during the production process.

# FOREWORD

This is an interesting book in two ways. First it provides an account of the extraordinary achievements in Babylonian astronomy as set out in a 400-line cuneiform text, MUL.APIN. Second, it presents a textual analysis to show that MUL.APIN is not merely a record of astronomical thinking of the period, but that it indicates how writing may itself have been instrumental in the advance of astronomical knowledge. In this way, it illuminates the much-debated relation between writing and science.

As the authors show, the astronomical knowledge expressed in MUL.APIN has many of the features we take as characteristic of science. It details lists of astronomical entities, stars, their relation to each other, their relation to the observer, to the seasons, to diurnal (night and day) events in the different seasons, and the calculation of leap years. The compilers of MUL.APIN even knew something that came as a bit of a surprise to me, namely, that the length of one's shadow is correlated with the season.

The authors cite an abstract formulation that appears in the latter portion of the treatise, described as an axiom: "*4 is the coefficient for the visibility of the Moon.*" They write: "This axiom...puts the astronomer scribes who wrote it well within reach of a formal, theoretical, mathematical science." But, as they note, the treatise also contains discourse of a decidedly non-scientific nature, the obligatory astrological implications pertaining not only to planting and harvest but also to the probable success of one's hopes and schemes.

The primary concern of the Watson and Horowitz book, however, is to explore the extent to which the advance of Mesopotamian astronomical science could have been, at least in part, a product of writing and literacy. There is no question that the science was built upon a long history of keeping records of times, distances, risings and settings, and measurements of angles and distances. But the authors speculate, further, that the very formulation of knowledge into the patterns, principles, and axioms that make up the text may reflect successive attempts by the ancient scribes to formulate written accounts that would be increasingly comprehensible to readers.

The component texts that comprise MUL.APIN indicate a progression over time, a reformulation of knowledge from simple lists of stars to expressions of complex relations amongst celestial and terrestrial events, to advancing definitions and drawing inferences. All of these are features that implicate, if not actually demonstrate, the uses of writing for science.

Two lines of work come to mind in relation to that presented in this volume. Chemla (2004) examines the role of writing in the evolution of science and mathematics in antiquity in several cultures, work that complements that of Watson and Horowitz. The second line of work that warrants comparison is Gladwin's (1970) celebrated work on Micronesian navigation. Gladwin studied the traditional, that is, preliterate, navigational practices still employed for sailing long distances out of the sight of land by the Caroline Islanders. The navigator memorizes the pattern of stars, comparable to the "star paths" described by Watson and Horowitz. The navigator then visualizes himself as the fixed centre of two moving frames of reference, one provided by the islands that eventually come into sight, the other provided by the pattern of stars which wheel overhead from east to west. What turns such sophisticated practical knowledge into science is the attempt to turn that practical knowledge into a form of a text that, as Watson and Horowitz show, is designed to be useful to a reader, shows reasoned progression, appeals to formalization and mathematization, and is useful for communicating and teaching knowledge.

This book is an important contribution to answering the question of just how writing something down could change our mental representation of it. Like Watson and Horowitz, I believe that it does, and continue to ponder just how.

David R. Olson
University Professor Emeritus
OISE/University of Toronto

# INTRODUCTION

This book presents the findings of an unusual collaboration, occasioned by a cuneiform tablet in the collection of the British Museum (BM 82671)[1] that was included some years ago in an exhibition on the history of writing. One of the significant features of the cuneiform text inscribed on the tablet is its organization: the lines are ordered by initial orthographic elements. To a developmental psychologist familiar with theories of literacy and cognition, the tablet was a minor revelation. Developmental research on orthographic awareness has focused primarily on print literacy and alphabetic orthographies, and nonalphabetic scripts are often assumed not to engender such awareness.[2] Yet orthographic elements clearly served as conceptual categories for the writers of this cuneiform text. What might the broader Mesopotamian cuneiform corpus suggest? Consultation seemed to be in order.

Rita Watson (RW) turned to Wayne Horowitz (WH), who was surprised by the question. As a traditional Assyriologist, most of his efforts had been focused on issues of text reconstruction, translation, and interpretation, as part of his ongoing study of the history, culture, and scientific tradition of the Ancient Near East,[3] and he knew that orthographic elements had influenced the organization of ancient cuneiform texts from the earliest exemplars (cf. Nissen, Damerow & Englund, 1993; Englund, 1998) to later forms that include the manipulation of signs in colophons (Hunger, 1968). *The Babylonian Theodicy*, a wisdom text, even had strict requirements on which syllabic sign would appear at the start of each line in eleven-line stanzas. Each line in the first stanza begins with the sign A: for the *a* of *anāku*, "I," starting

[1] A tablet from Girsu dated to ca. 2250 BCE that lists personal names beginning with the sign NIN; for an edition and discussion see Lambert 1988; for a cognitive perspective on the tablet, see Watson, 2000.

[2] See Harris, 2000:14, for discussion of the "alphabetic bias" in Western thinking.

[3] See, for example, the works of Neugebauer, Reiner, Sachs, Pingree, and their students and colleagues who straddle the realm of cuneiform studies and the history of science.

an acrostic whereby the author introduces himself by name and gives his profession.[4]

The discussion led to a cup of coffee and, eventually, to a collaborative journey that would be navigated from two very different points on the academic compass: an examination of cuneiform texts from a cognitive-analytic perspective. WH had not previously couched his observations in cognitive terms and was interested in discovering a new set of analytic tools with which to approach Mesopotamian cuneiform texts. RW was interested in what the cuneiform corpus might reveal about writing and conceptual change. The result was the exploratory analysis of the Mesopotamian astronomical treatise MUL.APIN presented in this volume.

The modern relevance of the ancient Mesopotamian astronomical tradition is illustrated in aspects of it that persist to the present day. We still measure time and space in accordance with the ancient Mesopotamian system. The division of a minute into sixty seconds and an hour into sixty minutes is a direct consequence of their sexagesimal (base sixty) mathematics. Our continued use of minutes and seconds of longitude and latitude ultimately derive from the Mesopotamian 360-degree geometric circle, which in turn can be related to their 360-day "ideal" year (Al-Rawi & Horowitz, 2001). MUL.APIN was a significant achievement within the Mesopotamian astronomical tradition, the culmination of which was the ACT corpus of Babylonian astronomical-mathematical material, which was transmitted to Greece and Rome.[5]

In Chapter 1, we describe MUL.APIN, its place in the Mesopotamian cuneiform text tradition, and why we chose it as the subject of this volume. Our analysis relies on the standard translation published by Hunger and Pingree (1989).[6] Making our own new translation could have had the undesirable effect of introducing our own biases into the text that we are proposing to study. Indeed, any act of translation can introduce a margin of error. However, as we could not assume a

[4] See Pearce, 1995:2275; Lambert, 1960:63 for Akkadian acrostics see Soll, 1988; Brug, 1990; and Hurowitz, 2002:331–332.

[5] ACT is an acronym for *Astronomical Cuneiform Texts* (cf. Neugebauer, 1955); for a summary of the *ACT* tradition see Hunger and Pingree 1999: 212–270.

[6] The Hunger and Pingree 1989 edition makes use of the 42 manuscripts of MUL.APIN that were known to the authors at the time of publication. Since then, a few additional fragments of MUL.APIN have been identified, but they add little or nothing to the reconstructed text.

general knowledge of the Akkadian language among our readers, and as our intention was to make the volume as widely accessible as possible, translation was necessary. In cases where translation seems to muddy the analysis, or where it fails to render fully the nuances of the original text, we draw on the original Akkadian text directly. We are fortunate, in that Hunger and Pingree's (1989) translation is highly faithful to the original. We are indebted to H. Hunger for his permission to quote freely from his study in this volume, and also, to add his and Pingree's English translation of MUL.APIN, *in toto*, as Appendix One.[7]

In Chapter 2, we discuss the issue of writing and conceptual change. The development of writing has been advanced as a possible explanation for the ascent of rationality and logic in the classical period, but this notion has been both challenged theoretically and underexplored empirically.[8] We also present a brief account of cuneiform literacy, its relevance to MUL.APIN and our analysis, and the importance of the cuneiform corpus to a broader understanding of the issue of writing and conceptual change.

Chapter 3 details the terms of the analysis presented in this volume and its cognitive-linguistic basis. A naturalistic perspective on cognition and language assumes that the universal biological endowment of human beings underwrites a meaningful degree of comparability, if not strict universality, in thought and language across cultures that may be disparate in both time and place. We here define the specific terms and categories that we apply to the MUL.APIN text.

Chapter 4 presents the text itself, along with the systematic application of the analytic categories and conventions of both fields, Assyriology and cognitive science; the text in its entirety, without annotation, appears in Appendix One. The results of the text analysis are summarized in Chapter 5 and discussed in Chapter 6. A cognitive perspective on writing and conceptual change in MUL.APIN is given in Chapter 7, and Chapter 8 offers a final word on how our analysis may serve to inform current understanding of MUL.APIN and its place in the cuneiform astronomical tradition.

---

[7] We also thank the publishers of Hunger and Pingree's edition, Eisenbrauns, for permission to reproduce the text.

[8] See Harris, 2009.

A collaborative endeavor such as this, that crosses borders between academic disciplines, clearly entails some risks. The rules of evidence and argument differ markedly between Assyriology and the cognitive sciences. However, we have endeavored to render our work comprehensible to any informed reader by making our assumptions and terminology as explicit as possible throughout.

A further difficulty is that of anachronism, since we apply contemporary theoretical understandings to an ancient text that is distinct linguistically, culturally, geographically, and temporally from those of our own era. However, we don't see this risk as insurmountable. The best of intentions cannot prevent cultural or historical bias, but we rely on the universalist assumptions of mind and language, outlined in Chapter 3, as a counterweight to biases which may linger, undetected, in this work.

There is an absence of precedents, procedural templates, and conventional criteria against which the significance of this unusual analysis can be easily evaluated, but it is our hope that it may illustrate how collaborative endeavors may yield new forms of understanding, and how diverse fields of inquiry may illuminate one another.

## *Conventions*

Assyriological abbreviations are as in *The Chicago Assyrian Dictionary* (CAD) with the exception of EAE = the series Enuma Anu Enlil. References to dates in the Assyriological material, e.g. 7th century library of Assurbanipal in Nineveh, are all BC/BCE. For the ancient Mesopotamian month names and their equivalents, see Appendix Two. The text includes cross-references, in which we refer forward or back to other sections of text that illustrate a particular point, or that contain related discussion. In this case, the first number of the cross-reference represents the chapter number and the following numbers, the section number. Thus, 4.3.1.2 refers to Chapter 4, section 4.3.1.2.

CHAPTER ONE

# MUL.APIN

## 1.1 *The Text*

MUL.APIN is a cuneiform astronomical treatise that appears in the early 7th century BCE. It is the first reasonably full exposition of the knowledge developed within the already centuries-old written tradition of cuneiform astronomical and astrological texts. The earliest of these date to the Old Babylonian times (c. 1700–1500), and include simple star lists in the tradition of the lexical series Urra = *hubullu*;[1] the oldest surviving mathematical astronomical work, a quantification of the change in the length of day and night over the course of the year;[2] and the earliest surviving astronomical omens in the tradition of the series Enuma Anu Enlil.[3]

Over the next fifteen hundred years, Mesopotamian astronomer-scribes developed an ever-more-sophisticated scientific astronomy. The last centuries of the second millennium give rise to the Astrolabe tradition.[4] At the start of the first millennium, the more advanced type of astronomy found in MUL.APIN appears, which subsequently gives way to increasingly advanced forms of astronomical endeavor from the end of the Neo-Assyrian period (7th c.). The final achievement of this extended tradition was the Babylonian astronomical-mathematical material found in the ACT corpus[5] of the Persian and Hellenistic periods, which was transmitted to Greece and Rome.[6]

[1] See Horowitz, 2005.

[2] Evidenced in tablet BM 17175+; see Hunger & Pingree, 1989:163–4.

[3] See Rochberg-Halton, 1982, for a fuller discussion of Enuma Anu Enlil.

[4] Horowitz, in press.

[5] See fn. 5, introduction.

[6] See Rochberg, 2008, for a detailed discussion of the Hellenistic transmission of Babylonian astral sciences; particularly pages 18–22 for the Greek awareness of the Babylonian inheritance; and Jones, 1999, for reference to "Orchenoi," or, people of Uruk, identified by Strabo as a group of "astronomical Chaldeans" (cf. Rochberg, 2008:18).

## 1.2 *Form*

In its standard form, MUL.APIN is written over two clay tablets and is comprised of almost 400 lines of text. It is divided into a number of sections and subsections, usually marked by horizontal dividing lines drawn by the ancient scribes. The subject matter ranges from simple star catalogues through detailed descriptions of lunar, solar, stellar, and planetary phenomena. The early sections (see Chapter 4, 4.1) are comprised of star catalogues, and illustrate one of the simplest extant written forms, a list structure (MA I i 1–9):

I i 1 ¶The Plow, Enlil, who goes at the front of the stars of Enlil.
I i 2 ¶The Wolf, the seeder of the Plow.
I i 3 ¶The Old Man, Enmešarra.
I i 4 The Crook, Gamlum.
I i 5 ¶The Great Twins, Lugalgirra and Meslamtaea.
I i 6 ¶The Little Twins, Alammuš and Nin-EZENxGUD (Gublaga).
I i 7 ¶The Crab, the seat of Anu.
I i 8 ¶The Lion, Latarak.
I i 9 ¶The star which stands in the breast of the Lion: the King.

Succeeding sections of text introduce more complex forms of expression (MA I iv 7–9):[7]

I iv 7 All these are the *ziqpu* stars in the path of the stars of Enlil which stand in the middle of the sky
I iv 8 opposite your breast, and by means of which you observe
I iv 9 the risings and settings of the stars at night.

Subsequent sections of MUL.APIN present a mix of observational science, measurements, and calculations. Yet the treatise also includes predictions of weather and human events, including omens, and astrological and mythological material that, to the modern reader, appears obscure (MA II i 26–31):

II i 26 on the day their stars become visible you observe their risings, their glow, and
II i 27 their…., and the wind that blows; you guard (?) the horses
II i 28 so that they do not drink water from the river.
II i 29 When their stars have been made visible,

[7] These entries constitute the summary ("conclusion") of the *ziqpu* star list, described in detail in Chapter 4.

II i 30 you present offerings to them; horses
II i 31 will touch bitumen and drink water from the river.

## 1.3 *Date of Composition*

The actual date of the final composition of MUL.APIN is uncertain, but the presence of numerous exemplars of the treatise in the library of the 7th-century Assyrian king Assurbanipal (668–627) demonstrates that the composition had reached its canonical form by this date. The slightly earlier colophon on an important source[8] for Tablet II of MUL.APIN dates this tablet to 687 and thus serves as a *terminus ante quem* for the canonical version of the treatise. Yet much of the content of MUL.APIN rests on earlier observations and calculation, such as the premise that the solstices are marked by 4:2 and 2:4 ratios for the longest to shortest days as measured on the water clock: this was known a millennium earlier than the earliest surviving dated copies of MUL.APIN.[9]

Other material in MUL.APIN is younger, but still hundreds of years older than the earliest dated copies of the treatise. The stellar sections at the beginning of MUL.APIN, for example, seem to be younger than similar material presented in Astrolabe B, a 12th-century compendium written in Assur.[10] The astronomical information embedded in the stellar sections of MUL.APIN is more accurate than that in Astrolabe B, indicating that MUL.APIN represents a later, improved state of astronomical knowledge. Likewise, there are indications, both within and outside the text, that the scribes who composed MUL. APIN understood that, in a sense, it updated the Astrolabe tradition.

The MUL.APIN text shows that the scribes knew about the earlier stellar system that stands at the heart of the Astrolabes,[11] where 36 stars rise in sequence, three stars per month, over an annual 360-day

[8] HH, an important source for Tablet II = VAT 9412, from Assur.

[9] This is found in the aforementioned Old Babylonian tablet BM 17175+ (Hunger & Pingree, 1989:163–4).

[10] For the the Astrolabes and their history, see Horowitz, 2007, and Casaburi, 2003. A new edition of Astrolabe B and the Astrolabe group by W. Horowitz is forthcoming. See also Horowitz 1998:154–166 and the early edition of Astrolabe B in Weidner 1915:62–102.

[11] See section b, Chapter 4, 4.2.

year circuit.[12] This same model of 36 stars is used in the astronomical text CT 33 9 from Assurbanipal's library in the middle of the 7th century, which has yielded numerous copies of MUL.APIN. The text CT 33 9 is written in the Astrolabe format, presenting 36 stars in the Paths of Anu, Enlil, and Ea, but the stars are distributed, by path, using the superior astronomical criteria of the type found in MUL. APIN (Horowitz, 1998:170). Thus, the date of Astrolabe B, the late 12th century, serves as a sort of *terminus pro quem* for the composition of MUL.APIN.

But was MUL.APIN written in a single burst of scientific creativity, once the required astronomical knowledge and techniques that underwrite its composition were in place, as early as 1200 BCE? Hunger and Pingree (1989:9–12; 1999:57) suggest not. They argue that MUL. APIN was "published" in its present canonical form not long before MUL.APIN's *terminus ante quem* of 687 BC, but suggest a date of ca. 1000 BC for much of the astronomical observations included in the series. If this is correct, then there is a window of at least a few centuries between the presumed date of the astronomical observations and our earliest exemplars of canonical MUL.APIN (7th century) for the series to have evolved into its canonical form.

Each of the various component sections of MUL.APIN[13] relates to a different set of observable or measurable astronomical phenomena. Each of these sections is, at least in theory, and to greater or lesser degrees, potentially autonomous. That is, individual sections may have had an independent existence, in some form and at some time, outside canonical MUL.APIN. Many, perhaps even most, of the component sections of MUL.APIN may have had their own history before being incorporated, with or without some light editing by the astronomer-scribes, into the canonical treatise. Other sections may have been composed specifically for MUL.APIN proximate to the "publication" of the canonical text. When viewed in historical context, then, the existence of parallels to various portions of MUL.APIN outside the series is significant (Hunger & Pingree 1989:9; George 1991). The consensus

[12] The Astrolabes give an ideal system that names one star that rose each month in each of the three paths of the Mesopotamian sky: the central path of Anu, the northern path of Enlil, and southern Path of Ea—for a total of three stars per month, and so 36 stars over the course of the year.

[13] Each component section is detailed in our analysis of the text, Chapter 4, and the overview of the sections using Hunger & Pingree's divisions can be found in Appendix Three.

position, that MUL.APIN was assembled from parallels to its component parts, requires exactly this type of material to exist outside of MUL.APIN.

It must be emphasized that no direct evidence for such a process is extant. We have only the canonical form of the treatise, from the time of its earliest exemplars in the 7th century down into the Hellenistic period. No "forerunners," or specific "pre-MUL.APIN" group of texts that can be unequivocally identified as precursors survive. Thus there is no way to determine precisely the historical process by which MUL. APIN was composed.

There are, however, parallel historical-developmental processes in the Enuma Anu Enlil tradition, where numerous second millennium predecessors to the first millennium series *are* extant (Hunger & Pingree 1999:7–12; cf. also Rochberg-Halton, 1982), and in the Astrolabe group, where component parts of Astrolabe B exist independently both *before* and *after* the time of Astrolabe B (Horowitz 2007:107–108).

Also, the astronomical fragment K.7931, from the 7th century library of Assurbanipal at Nineveh, may show evidence for processes of this type. This fragment[14] bears a colophon which identifies it as DUB.ME LIBIR.RA.ME GABA.RI KÁ.DINGIR.RA[ki], "copied from older tablets from Babylon."[15] The tablet itself is an anthology, consisting of four sections, all of which include star names. Thus, the sources for K.7931 must be first, astrological in nature; second, older than K.7931; third, must have come to Assyria from Babylonia; and fourth, must have been edited in Assyria, in some way, before being included in K.7931.

Unfortunately, the first three sections of this tablet are too badly damaged to be identified, so we cannot make full use of this fragment to substantiate the forgoing point regarding the transmission and composition of canonical MUL.APIN. However, it is clear that the last section of K.7931 parallels, simultaneously, both the Astrolabe tradition and Enuma Anu Enlil Tablet 51—the tablet of the series that gives omens relating to Astrolabe stars, but with only one star selected for each month, instead of three stars per month as in the main Astrolabe

---

[14] The fragment will be published in full in Horowitz, forthcoming.

[15] Hunger Kolophon, Hunger, 1968:95–6, no. 312, Bezold Cat. II 883. Hunger restores the colophon to identify the tablet with the scribe of a number of astronomical and astrological tablets *Ištar-šumu-ēreš* grandson of *Nabû-zuqup-kēnu.*

Table 1. A Comparison of K.7931 Section 4 and the Omens in Enuma Anu Enlil 51

| Star | K.7931 Section 4 Rising month | Enuma Anu Enlil 51 Rising month |
|---|---|---|
| MUL.MUL | II | II |
| $^{mul}$KAK.SI.SÁ | IV | IV |
| $^{mul}$BIR | VI | VI |
| $^{mul}$GÍR.TAB | VIII | VIII |
| $^{mul}$TI$_8$ | X | X |
| $^{mul}$KU$_6$ | XII | XII |

tradition (See table 1).[16] Thus, K. 7931 not only represents a secondary use of the older Astrolabe repertoire of stars, but further demonstrates that this set of stars was used and reused. It is this type of reuse and reorganization of earlier source material that we suggest is implicated in the development of canonical MUL.APIN.

### 1.4 *MUL.APIN and the Scribal Tradition*

The assumption of preservation and transmission of the content of MUL.APIN rests in large part, of course, on what is well known about the cuneiform scribal tradition.[17] Some texts survived in recognizable form for thousands of years within this tradition. Preservation was influenced in part by educational institutions, within which the copying of texts was the primary mode of instruction, but also by a concern for the preservation of older texts and veneration for the knowledge that they contained (Nissen, Damerow, & Englund, 1993; Veldhuis, 1997, 2004; Pearce, 1995). That is, while scribal education was dedicated to the teaching of specific skills, the central concern of the curriculum may be best understood as an induction into the scribal tradition, and the cultural heritage and knowledge contained therein (Veldhuis, 1997:82–3).

[16] Tablet 51 is edited in Reiner & Pingree, 1981:52–69 with further discussion in Horowitz, forthcoming.

[17] The scribal tradition is further discussed in relation to cultural transmission in Chapter 2, 2.1, this volume.

The content of canonical MUL.APIN is known to have been studied and repeatedly copied by astronomer-scribes across generations, from the time of its appearance in the 7th century down into the Hellenistic period. We can assume, based on the foregoing, that the knowledge contained within the component sections of MUL.APIN was transmitted within the scribal tradition and, as Hunger and Pingree surmise, MUL.APIN does not reflect the knowledge of a single scribe who composed the treatise in one sitting near the time of its "publication." Rather, it represents the cumulative knowledge held in common by the entire community of Mesopotamian scribes who used and copied it.

The score of transcripts of MUL.APIN included in Hunger and Pingree's (1989) edition show little variation from manuscript to manuscript. This is consistent with the conventions of the scribal tradition, within which the text would have been passed from generation to generation, almost unchanged, from the Neo-Assyrian period down to Hellenistic times. It is thus highly unlikely that the sequence of component sections of MUL.APIN is random, or the result of editorial or historical accident. The structure of the canonical version suggests, rather, a long period of incubation. Neither the intentions of the compilers, nor the method they used to create the canonical version of MUL.APIN, can be verified. The scribes left us no reflective comments on the process in which they were involved. The content and form of the text itself, however, may yield some clues as to the way in which it evolved.

## 1.5 *Sequence in MUL.APIN*

There is a clear tendency for older types of information and text forms to appear early in the canonical MUL.APIN treatise, while later-emerging astronomical content and text forms appear later in the treatise. For example, the older knowledge of the type found in the opening sections a–d of MUL.APIN, provide the type of information that is available in an earlier, less precise, form in Astrolabe B.

More specifically, the first section, MUL.APIN a, gives a star catalogue of the type found in Astrolabe B, Section II, with its 36 month-stars (see fn. 10, above). Similarly, the dates for the rising of stars, and simultaneous rising and setting of stars, found in MUL.APIN sections b–d, parallel Astrolabe B, sections III–IV. In these cases, the astronomical data in MUL.APIN is more exact than that found in the older, parallel Astrolabe texts. This is not surprising, even were we to ignore the fact that MUL.APIN is later than Astrolabe B. Calendrical, as well as religious and astronomical, considerations condition the astronomy of the Astrolabes. For example, the month of Kislev (Month IX) in the Astrolabe tradition is identified both with the planet Mars and the deity associated with this planet, Nergal, the King of the Underworld.[18] Further, Astrolabe B is bound by its calendrical-astronomical theory, to assign one star to each path *per* calendar month. No such restrictions are placed on the editors of the stellar sections of MUL.APIN, Tablet I. After sections a–d, MUL.APIN continues with topics beyond the scope of the Astrolabes.

### 1.5.1 *Sequence: Procedural Considerations*

Another factor that may influence sequence in MUL.APIN is the level of technical expertise required to carry out the relevant astronomical observations and calculations. The series of component sections of the treatise refer, either explicitly or implicitly, to a range of procedures of increasing complexity.

The early component sections of MUL.APIN require only the most simple of technical expertise, similar to that required by the Astrolabes: naked eye astronomy and knowledge of the night sky. In section a, the single-entry format of the star-catalogues implies that the user of the text will observe only one star at a time, or at most, that star's position relative to its neighbor's. Similarly, in section b the reader of the text need only correlate a series of individual stars that rise in sequence over the course of the year with a set of given dates.

In sections c–d, the procedures implied are more complicated. These sections detail the simultaneous rising and setting of stars. Two stars must be observed at the same time, along with the date of the observation added in section d.

In the succeeding sections, the technical knowledge required by the "observer of the sky" (the astronomer-scribe making use of

[18] For some of the numerous examples of Nergal = Mars see most recently RlA 9 222–223, Reynolds 1998:353–354, and Koch-Westenholz 1995:128–129.

MUL.APIN)[19] becomes increasingly complex, as is evident in section e, which presents the *ziqpu* stars. At this point in the text, the astronomer is required to compare three sets of stars: those rising, those setting, and those reaching their culminations at the *ziqpu* point (the highest point above the horizon that the *ziqpu* stars were assumed to reach). Now, the "observer of the sky" must correlate the observations specified in sections c–d with observations of the stars overhead. After section e, MUL.APIN progresses to the subject of the Moon, then the Planets, and then the Sun. The final sections of Tablet II assume the skilled use of technical apparatus, the gnomon and the water clock.

The sequence of component sections in MUL.APIN thus parallels the history of ancient Mesopotamian astronomical texts, starting with lists and ending with procedural instructions. The succeeding sections also require increasingly sophisticated procedural knowledge that reflects ongoing improvements in astronomical technique, a progression that culminates with introduction of the technical apparatus, the gnomon and the water clock, used to measure time.

In summary, then, while the specific processes by which the canonical form of MUL.APIN evolved cannot be reconstructed with absolute certainty from the extant textual evidence, it is highly unlikely that the sequence of the component sections in MUL.APIN is random, or the result of editorial or historical accident. Historical considerations would suggest that the astronomers knew, by oral tradition or from written materials no longer extant, which components belonged first, presumably because they were deemed to have been written earliest, or to reflect older knowledge. Procedural requirements could also have influenced the sequence of component texts, since the material presented bears an ordered relation to the series of astronomical concerns, procedures, and calculations described. Each succeeding section of the treatise demands knowledge of previous sections, much as one would find in an academic textbook.[20]

The sequence of component texts follows what must have seemed to be a logical sequence to the astronomer-scribes who composed MUL. APIN. Current empirical research on text understanding also suggests

---

[19] See our discussion of the forms of direct address that appear for the first time in MUL.APIN in section e, the *ziqpu* stars, Chapter 4, 4.1.4.2; both second- and third-person nominal and pronominal forms are used.

[20] For an analogous genre, see 8.1, this volume, on the Mesopotamian Technical Handbook tradition.

that sequence in texts usually reflects an assumption of chronological order (Kintsch, 1998; Zwaan, 1999). This would be consistent with MUL.APIN as we understand it.

## 1.6 *Mesopotamians and Moderns*

How can we be sure what "assumptions" the astronomer-scribes had in mind? How did the scribal tradition[21] work, and how was knowledge passed on through succeeding generations? Clearly, texts were central, but in a large practicing community of scribes, oral tradition would have played a significant role. Omissions from texts most likely reflected what the scribes were assumed to know (Nissen, Damerow, & Englund, 1993:20). But to what extent can we infer backward in historical time? Can current thinking inform our understanding of a text composed thousands of years ago, by Mesopotamian astronomers living in an era so far removed from our own?

Perhaps the gap is not as great as it seems on first glance. The period of twenty-seven hundred years or so that separate us from the Mesopotamians who composed MUL.APIN is the blink of an eye in evolutionary terms. It is, in fact, a shorter span of time than that from the first cuneiform tablets to the last (Walker, 1990). Most extant accounts of the origin and evolution of human cognition contend that there have been no substantial changes in human cognitive endowment for at least forty thousand years (Holden, 1998), and on some accounts, much longer (Tomasello, 1999).[22] There are thus no grounds for assuming that the inhabitants of ancient Sumer and Akkad were different, biologically or cognitively, than modern humans. How, then, can we account for the very evident differences between Mesopotamians and moderns?

The MUL.APIN treatise illustrates that Mesopotamian astronomy, and assumptions regarding its proper domain, diverge significantly from those of modern-day astronomy. An example of the difference is the Mesopotamian sense of what we call the atmosphere and sky. On our view, the *atmosphere* ends where the domains of space, or outer space, begin. Solar and stellar phenomena belong to the latter. For

---

[21] The cuneiform scribal tradition is discussed in more detail in Chapter 2, section 2.1, this volume.

[22] Cognitive evolution and cultural transmission are more fully discussed in Chapter 2, 2.2, this volume.

Ancient Mesopotamians there was no sense of atmosphere since they had no knowledge of outer space. Instead, they assumed that air filled the entire universe, including both the heavens above and underworld below. This would also seem to be the case in traditional western Judeo-Christian views of heaven and hell, whose inhabitants require no special breathing apparatus. On the other hand, our common-sense term "sky" includes the near regions where birds fly as well as more distant regions where the Sun, Moon, and stars can be observed. Only rarely, and usually in a poetic sense, can "sky" refer to the realm of God above.[23]

Yet the sky itself has changed only slightly over the millennia separating ourselves from the MUL.APIN astronomers, changes defined primarily by the dates of stellar phenomena, and small changes in the shape and configuration of constellations. The differences we find between MUL.APIN and modern astronomy must therefore be attributed primarily to cultural variation, rather than to variation in the physical phenomena being described. That is, in spite of the variation between our culture and theirs, the domain of Mesopotamian astronomy remains overwhelmingly recognizable to us as astronomy. Some of its associated principles of measurement, as we pointed out in the introduction to this volume, survive intact to the present day.

From a cognitive-evolutionary perspective, the linguistic and cognitive endowment of Mesopotamians in the era of MUL.APIN would have been essentially the same as modern-day humans. They spoke a different language, or rather several different languages, and wrote on clay tablets using a script very different from our own. But these cultural variations occur against a background of essential similarity.

In short, Mesopotamians were like us. No other conclusion is consistent with any extant view of the nature and origins of human cognition. Our analysis of MUL.APIN in this volume is grounded on these assumptions. This "naturalistic approach," detailed in the next two chapters, is neither reductionist nor deterministic, nor does it supplant textual or rhetorical analysis. Some portions of the MUL.APIN treatise, indeed, require a rhetorical analysis in order to be understood. But a naturalistic approach has the distinct advantage of allowing an

[23] For an interesting variation on this theme in the classification of the animal kingdom of ancient Mesopotamia, see P. Wapnish, (1985) *Animal names and animal classification in Mesopotamia: An interdisciplinary approach based on folk taxonomy*. PhD Dissertation, Columbia University.

appeal to objective evidence on human cognitive and linguistic abilities, their evolution, and their biological bases.

Understanding an ancient text such as MUL.APIN requires some reconstruction of its interpretive context. A naturalistic approach puts tighter constraints on that reconstruction than a textual or rhetorical approach alone because it aligns itself with available empirical evidence. In so doing, it lends validity to speculation about the assumptions that the ancient authors and compilers of MUL.APIN may have had in mind.

### 1.7 *Analytic Considerations: Why We Chose MUL.APIN*

On our view, the MUL.APIN treatise provides the best available "test case," perhaps within the entire currently extant cuneiform corpus, for a cognitive-linguistic analysis. The text of MUL.APIN remained stable over hundreds of years, with very little variation among the surviving exemplars (Hunger & Pingree, 1989; 1999). So, for example, Source K from the Hellenistic period, during the reign of one of the Seleucid kings, consists of the same text as that of the much earlier Neo-Assyrian period with only minor variations in sign selection such as the use of the sign MÚL () instead of MUL (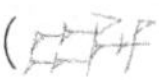) for the star-determinative, as is typical of Hellenistic period astronomical works. There are also the occasional variants, such as writing, in syllabic Akkadian, *e-la-ma-tú* in MUL.APIN I ii7 for *elammattu* (Elamite), instead of writing NIM.MA-$tu_4$ making use of logographs from Sumerian.[24]

Moreover, we have an authoritative modern edition of the text of MUL.APIN (Hunger & Pingree, 1989), with translation and transliteration, which can serve as a basis for our analysis. This state of affairs, a stable, nearly complete ancient text with a reliable modern edition stands in marked contrast to the majority of the cuneiform corpus, in particular the two other main groups of traditional astronomical/astrological texts: the series Enuma Anu Enlil (EAE) and the Astrolabe group.[25] For EAE, nothing resembling a modern edition is possible:

[24] In Source A in Neo-Bablyonian script dating to ca. 500 BC and Source FF in Neo-Assyrian script, which must pre-date the fall of the Assyrian Empire at the end of the 7th century.

[25] For an overview of Enuma Anu Enlil see Hunger & Pingree 1999:12–20. For the Astrolabes, see Horowitz, 2007. By "traditional," we mean texts from before the end

> Due to the bad state of preservation of most of the exemplars (of Enuma Anu Enlil)[26] and the lack of a complete text edition, a reasonably comprehensive overview of the series cannot be given.
>
> Hunger & Pingree, 1999:14

In addition, it appears that EAE never coalesced into a single, universally recognized form. Much work still needs to be done on the surviving manuscripts of EAE before we will be able to answer even simple questions regarding the redaction of the series, such as whether there was only one standard edition of EAE during the first millennium, or whether competing editions circulated (Hunger & Pingree, 1999; Koch-Westenholtz, 1995:78–79).

For the *Astrolabes*, we have a different problem: the group never reached a canonical form which could be passed down from generation to generation. Hence, the four sections of the earliest and most complete form of the *Astrolabes*, the so-called 12th-century *Berlin Astrolabe*, better known as Astrolabe B, never occur together on any earlier or later tablet belonging to the group, although each of the four sections survives separately into the first millennium (Horowitz, 2007).

Similar problems arise in the rest of the cuneiform corpus. Catalogues of ancient cuneiform texts include numerous works which remain unknown to us (Lambert, 1962) and reconstruction of standard editions of even the most well-known texts are still hindered by the incomplete nature of the archeological record. Even the most famous works of Akkadian literature, the Gilgamesh Epic (George, 2003), Enuma Elish and "The Babylonian Job," *Ludlul Bel Nemeqi*, cannot yet be restored in their entirety. Such difficulties are endemic to the study of the cuneiform corpus where authoritative texts can only be established on the basis of tablets recovered from the remains of long-dead cities, archives, and libraries.

Thus, the existence of something approaching a fully reconstructed authoritative text for MUL.APIN is a welcome exception to the general "state of play" in Ancient Mesopotamian cuneiform scholarship. The stability of the canonical form of MUL.APIN over centuries makes it particularly well suited to be a test case in the application of a cognitive-linguistic analysis to an Ancient Near Eastern scientific text.

---

of the Neo-Babylonian period, when new techniques and text types come to dominate Babylonian astronomy and astrology, including the ACT tradition for mathematical astronomy, the zodiac, and the horoscope.

[26] Brackets, ours.

Multiple editions represented by incomplete manuscripts of a series that continued to evolve would have made our task next to impossible.

## 1.8 *Conclusion*

MUL.APIN seems, at least on first glance, to stand in isolation. No surviving written precursor to the series is extant, and as a result we have no *direct* evidence for the manner in which the astronomer-scribes composed canonical MUL.APIN. Similarly, direct "descendants," or subsequent versions of the series have yet to be discovered. If such descendants were extant, and if they diverged from the canonical form of the treatise, they might have provided us with some insight into how the text ultimately came to be used.

In the absence of these variant textual forms, an examination of "internal" evidence, the content of the text itself, is particularly useful. A comparison of MUL.APIN with contemporary and earlier cuneiform astronomical works (Hunger & Pingree, 1989:10–12) may be the only way at present of gaining insight into how and when the text may have been composed.[27] The analysis that we present in the current volume is an additional way of gaining insight into the text, and thus of addressing questions that remain unanswered by the archeological record. The results of our cognitive-linguistic analysis may or may not support Hunger and Pingree's (1999) conjectures, but either way, we will be one step closer to a fuller understanding of the text.

[27] The situation for MUL.APIN may be compared to that of Homer in the classical world, where an authoritative text is assumed to be the result of long-term redaction of a text tradition but for which no manuscripts survive; for Assyriology, a comparison can perhaps be made to Enuma Elish, which was composed in the late 12th century, incorporating pre-existing content and traditions, but for which no manuscripts earlier than the first millennium survive; another comparison, to Jewish sources, might liken the study of the text history and redaction of MUL.APIN to the written Torah, while that of Enuma Anu Enlil in its various forms requires a methodology similar to that of the Talmud.

# CHAPTER TWO

## WRITING AND CONCEPTUAL CHANGE

Language is a universal human faculty[1] but literacy is not. Writing is a relatively recent cultural invention that first appeared, as we know it, in the civilizations of the Fertile Crescent in Mesopotamia in the late fourth millennium.[2] It is not known whether Sumerian cuneiform writing was an impetus for the development of Egyptian systems, and it is probable that they developed independently, as did the Chinese and Mayan writing systems (Michalowski, 1994:53). In all cases, the emergence of writing appears linked to increasing social, political, and economic complexity, and a concomitant requirement to administer and control the transfer of commodities (Cooper, 2004:72; see also Nissen, Damerow & Englund, 1993): settled populations with agrarian economies produced a surplus of goods, which resulted in an increase in trade. However, as Cooper (2004:94) has noted, literacy is not an obligatory marker for complex societies or civilizations.

The cuneiform corpus provides a uniquely rich record of the early uses to which writing was put, largely because of the durable material on which it was written: clay tends to outlast papyrus and other, softer materials. The necessity for record keeping and the relative stability brought by urbanization allowed both the preservation of written records and the elaboration and development of the writing system. From the earliest texts of the archaic period we can observe how written texts evolved within the literate tradition of the cuneiform scribes.

---

[1] See below, 2.2.3, and Chapter 3, this volume, for discussion.

[2] For accounts of writing, see Gaur, 1987; Gelb, 1963; Harris, 1986, 2000; for a survey of different systems, Daniels & Bright, 1996; for discussions of writing in relation to cultural and conceptual change, see Goody & Watt, 1968, Goody, 1977; Olson, 1977, 1994; 2001; Ong, 1982; Stock, 1983.

## 2.1 *The Cuneiform Scribal Tradition*

The cuneiform scribal tradition lasted more than three millennia, around thirty-four centuries. Some of the lexical series survived in recognizable form for thousands of years within this tradition.[3] Through the literate practices associated with their profession, scribes preserved the cultural heritage of Mesopotamia and the accumulated knowledge of that culture. The stability of textual forms and content across generations of scribes and centuries of tradition has been widely attested,[4] and several factors may have influenced this.

The institutions of scribal training, such as the scribal school or edubba, were a major stabilizing factor, as the copying of texts was the primary mode of instruction. A conscious concern for the preservation of older texts, which the scribes recognized as containing important academic knowledge, also fostered conservatism and traditionalism (Veldhuis, 1997:80), as did the high regard in which the Sumerian language and traditions were held. For millennia after it had ceased to be a spoken language, Sumerian maintained a status akin to that of Latin in the Western intellectual tradition (Nissen, Damerow & Englund, 1993). The elevated social status and important administrative role of the scribes also led to a sense of exclusivity being associated with the scribal tradition.

Obtaining a scribal education was no small accomplishment, and the difficulties inherent in mastering the cuneiform system, with its complex signs and multiple phonetic readings, ensured a lengthy period of scribal education (Pearce, 1995:2270). Scribal education was central to the administration of the state, but it is not clear whether influence was exerted on scribal education by a central authority or whether priestly families and smaller schools were more important in this regard.[5] Finally, and of primary concern to our discussion in this chapter, writing itself was a stabilizing factor because of the permanence of representation that it allowed.

---

[3] For accounts of cuneiform writing, see Walker, 1990; Nissen, Damerow, & Englund, 1993; Vanstiphout, 1995; Black & Tait, 1995; Veldhuis, 1997; 2004; Civil, 1995, 2000; Michalowski, 1994, 1995; Cooper, 2004; Houston, 2004.

[4] Note, however, that there were periods of change; see Civil, 1995, and also Veldhuis, 1997.

[5] See Veldhuis, 1997:25–28, 79–83, for a detailed discussion of factors governing stability and change, and the transmission of the scribal curriculum; see also Rochberg, 2008.

It should be noted that the *stability* of textual form and content within the scribal tradition is not tantamount to exact *replication*: changes occurred in the process of transmission.[6] The writing system itself underwent substantial development over the centuries (Walker, 1990; Vanstiphout, 1995; Veldhuis, 1997). It would have been impossible for changes *not* to occur in the texts. Also, dynasties came and went, and administrative centers moved. Along with these shifts in the social and political sphere, different cultures and spoken languages emerged as dominant. Scripts had to be adapted to represent them.[7]

The complexities of adapting the cuneiform script to more than one language, and the lengthy process of teaching associated scribal techniques, led to a formal analysis of written language that was arguably unequalled in its sophistication until the linguistic science of the modern period.[8]

### 2.1.1 *The Cuneiform Lists and Conceptions of Language*

The earliest extant written form is the list. Early cuneiform lists were central to scribal education, and written language is a clear factor in their organization. Some lists are arranged according to the stylus strokes needed to write the signs they contain, which would seem to serve a clear pedagogical purpose. List content ranges across the scribal curriculum,[9] but a formal analysis yields a clear picture of types within the cuneiform corpus.

---

[6] See also the discussion of cultural transmission, below, 2.2.5.

[7] See Rubio, 2007, for a detailed description of "alloglottography."

[8] See discussions in Civil, 1995; 2000; Vanstiphout, 1995; and Veldhuis, 1997. Harris (2009) notes that the development of modern linguistics awaited the work of de Saussure; although see Robins, 1997, on Stoic linguistics, and Harris, 2009:137–8. Aristotle's notion of definition was linked more to epistemology than to linguistics: a definition was the "essence" of the thing, and the essence of a thing was its definition, although a literate concept of "word" may have figured in this development (Watson, 1985, 1995; Watson & Olson, 1987).

[9] The curriculum, as attested to by extant scribal texts from various periods, included sign lists, vocabularies, syllabaries, mathematics, accounting, and divination; mathematical tables for multiplication, reciprocals, squares and square roots, coefficient lists presenting fixed values for various items, problem texts in algebra, geometry, and surveying (Pearce, 1995:2271); see Veldhuis, 1997, for a detailed account of scribal education.

> (there are) lists organized according to shape, or some other characteristic of the cuneiform signs (or logograms); lists organized thematically, according to the meaning of the logograms; lists organized phonologically, following the reading of the logograms; etymological lists; synonym lists, and miscellaneous lists.
>
> Civil, 1995:2309

The extent to which writing influences the organization of the lists is evident from Civil's typology. However, writing is not the sole organizing factor. Thematic lists, such as the encyclopedic Urra = *hubullu* series, organize logograms according to their meaning, or related conceptual content: animals, trees, stones, metal objects, wooden implements, place names, divine names, and so on (Civil, 1995:2311). The list of professions from the archaic period can even be considered a "sociological" picture of the society of the time (Nissen, Damerow & Englund, 1993).

Writing, then, did not initially determine the content and structure of lists, but very early on it seems to have "intruded," and thereafter became a dominant organizing principle in many of the texts that survive. That is, writing served not only to *represent* content but also to *organize* that content in particular ways. The pedagogical goals of the scribal tradition were undoubtedly a factor in the form and content of the lists, since scribes had to learn the signs they would be called upon to use in the practice of their profession. But the script itself had an effect. Cuneiform orthography clearly influenced the manner in which the scribes organized the lists.

## 2.2 *Writing, Cognition, and Culture*

Becoming literate entails the development of new conceptual categories. At the very least, learning how to use an orthography requires learning correspondences between the script and the syllabic, phonemic, or semantic constituents of a language that the script may be used to represent. Such learned correspondences become part of a literate individual's cognitive repertoire, as has been empirically demonstrated.

Alphabetic literacy requires the formation of phoneme-grapheme correspondences. Empirical research carried out with children shows that learning to read an alphabetic script is related to a reliable and measurable increase in phonological analytic abilities.[10] These abilities influence the way in which spoken language is perceived, and they are specific to the type of orthography being learned. They do not manifest in logographic literates (Read, Zhang, Nie, & Ding, 1986) nor is there any reason why they should. Acquiring a logographic script does not require the development of phoneme-grapheme correspondences. The organizing principle of logographic scripts is neither strictly phonemic nor strictly conceptual, or "ideographic." Semantic and phonemic elements appear in combination; and logographic characters take a more complex description than letters of the alphabet.

### 2.2.1 *Literacy and the Brain*

Learning to read changes the functional organization of the adult brain. In one study, newly literate and non-literate adults of the same age, and with identical socio-cultural backgrounds, were asked to analyze spoken, unfamiliar pseudo-words (Castro-Caldes, Peterson, Reis, Stone-Elander & Ingvar, 1998; cf. Frith, 1998). This task requires the kind of analysis necessary to the acquisition of alphabetic literacy: grapheme-phoneme correspondences. Graphemes are visual representations, in this case, written alphabetic signs. Learning an alphabet entails learning how these graphemes correspond to a system of sounds, or phonemes: the phonology of spoken language. When literate and non-literate adults performed identical phonological discrimination tasks in this study, the pattern of brain activation was found to differ between them.

The difference between alphabetic and logographic orthographies also manifests at the neurological level. Dyslexia patterns differently in the brain depending on the orthography acquired (Tan, Laird, Li & Fox, 2005; Siok, Perfetti, Jin & Tan, 2004). Reading-impaired logographic literates manifest deficits in mapping correspondences between graphemes and syllables, and also between graphemes and concepts; both processes mediated by the left middle frontal gyrus (Siok et al., 2004).

---

[10] See Alegria, Pignot, & Morais, 1982; Bertelson & DeGelder, 1994; Bradley & Bryant, 1983; Castles & Coltheart, 2004; Cheung, Chen, Lai, Wong & Hills, 2001; Martinet, Valdois & Fayol, 2004; Shankweiler & Liberman, 1972; for a comprehensive review of reading, see Stanovich, 2000.

This is in contrast to the deficit in mapping correspondences between graphemes and phonemes that manifests in alphabetic literates, a process associated with the left tempo-parietal region of the brain.

Different orthographies, then, not only have different representational properties, but also have different associated neurological effects. Writing is a cultural invention, but it has measurable effects that manifest at both neurological and cognitive levels. Does biology, then, have a place in cultural explanation?

### 2.2.2 *Naturalistic Approaches*

The cognitive turn in the social sciences, around the mid-twentieth century, brought with it a search for universal biological explanation. Answers were sought in nature rather than experience (cf. Pinker, 2002), and interest in cultural explanation waned. But in the last few decades this has changed. Culture has been found to affect not only cognition, but also biology. Naturalistic approaches seek to ground cultural explanation in the biological substrates of human cognition.[11]

On a naturalistic account, human mental abilities make culture possible and must, to some meaningful extent, influence its content and organization. Thus, even allowing for cultural variation, the basic genetic endowment that is common to all human beings plays a role in explanation. On most naturalistic accounts, brain states *are* mind states. While Descartes maintained that there exists a fundamental distinction between mind and body, methodological naturalism rejects Cartesian dualism.[12] Neurological activity is generally held to indicate corresponding cognitive activity, and neurological effects reflect corresponding cognitive effects.

One need not be a methodological naturalist, however, to recognize the brain bases of behavior. Experiences, thoughts, feelings, and emotions all correspond with physiological activity in the tissue of the brain (Pinker, 2004). At the very least, brain states can be assumed to be *related* to cognitive states, and even on this weaker claim, the neurological effects of literacy can be taken as strong evidence for corresponding

[11] For naturalistic epistemology, see Kornblith, 1985/1994; cognitive accounts can be found in Atran, Medin, & Ross, 2005; Sperber, 1996; Sperber & Hirschfeld, 2004.

[12] See Antony & Hornstein, 2003, and Pinker, 2002, 2004 for related discussions.

cognitive effects, as the correspondence between brain states and mind states is no longer in doubt. Acquiring the uses of orthography influences both the brain and cognition. Reading and writing, whether one employs an alphabetic, logographic, syllabic, or mixed orthography, can be expected to have cognitive effects.

Cultural explanation increasingly relies on biological metaphors, as the underlying causal elements in human behavior are generally held to have a biological substrate. Indeed, biology is far more than a metaphor in current cognitive science. However, it must be emphasized that a claim for a biological substrate to cognition is not equivalent to a claim that human cognition and behavior are *determined* by biology. That is, arguing from the biological to the cognitive is not a form of biological determinism (Pinker, 2002, 2004), nor is it incompatible with cultural explanation. Human beings are organisms motivated by internal states, not objects driven by physical forces, nor animals driven by unexamined instincts. Beliefs and intentions are primary causal factors in human cognition and behavior. Planning and reflection, and flexible higher-order cognition may operate over the results of peripheral or basic cognitive processing without being determined by them (cf. Carruthers, 2002; Dennett, 1994). Cultures are the cumulative products of such higher-order cognitive activity. Cultural evolution, on a Darwinian analogy, and the epidemiology of representations are two sorts of theories that draw on biological metaphors to explain the spread of ideas through populations of minds, across generations.

### 2.2.3 *Cognitive Evolution*

All human beings share a common genetic endowment, and some aspect of that endowment must be unique to the human species. There is no other explanation for why human beings, placed in the same environments as other species, follow such different ontogenetic paths. Current theories of the origins and evolution of human cognition all attempt to account for this, albeit in different ways.[13] Some views lean toward a more deterministic account, in which distinct

---

[13] See Boyd & Richerson, 1985; Dennet, 1999; Hauser, Chomsky & Fitch, 2002; Pinker & Jackendoff, 2005; Richerson & Boyd, 2005; Sperber & Claudiere, 2006; Sperber & Hirschfeld, 2004; Tomasello, 1999 for discussion and varying views.

human abilities are a consequence of particular genetic endowment. Genetically defined parameters, or constraints, specify a range of possible cognitive contents and structures, enabling rapid and successful development. This argument is usually made from language: many species communicate, but only human languages have the property of recursion (Hauser, Chomsky & Fitch, 2002; Pinker, 1994). A syntactically-complex, productive grammar enables human beings to create and understand an infinite number of unique sentences. By extension, modular views of cognition account for thought in a similar way (Fodor, 1983; 2001).

In spite of differences between languages and cultures, human thoughts are similar enough to enable their expression and comprehension in a shared language, and also to allow coherent translation between diverse languages. Parameters and constraints that underlie universal abilities, then, may take different values in different linguistic environments, but underlying competence is assumed to be universal.

Much of the criticism directed at this view has mistakenly attributed to it a biological determinism that extends to the realm of higher cognitive functions, such as beliefs, intentions, free will, moral understanding and the like, but this is not the case (Pinker, 2002, 2004). Constraints, on universalist accounts, describe *enabling* functions rather than *limiting* functions. Like the training wheels on a bicycle, they enhance the survival of the species by lessening the likelihood of error. At the level of basic cognition, the need to identify objects, predict actions, and thereby navigate the world of daily life, is rendered more effective and less error-prone by innate tendencies to draw the right sorts of inferences.

Other views of innate endowment contend that human uniqueness is a consequence of highly-evolved social cognition that enables learning through engagement with others (Tomasello, 1999; 2003). This argument is usually made from cultural learning. Other species are social, and can be argued even to have culture of a sort (cf. Kuscaj & Highfill, 2005), but only humans accumulate knowledge and transmit it across generations (Tomasello, 1999; Tomasello, Carpenter & Liszkowski, 2007). Other species benefit from cooperation, but only humans benefit from the cumulative, stored consequences of cooperation; and symbolic functioning and intentional understanding are substantially more developed in human cognition (Tomasello, 1999; Tomasello et al. 2007). The flexibility, planning, and higher-order cognition that differentiate human cognition from that of other species have been argued to be a consequence of language (Dennett, 1994).[14]

A point of contention is whether the difference between humans and other species is one of kind or of degree. The map of human uniqueness is being continually redrawn at present, as new empirical findings are made. The descended larynx, for example, long thought to be the unique basis of human speech, has recently been found in other species (Fitch & Reby, 2001; Fitch, 2007).

### 2.2.4 *Cultural Variation*

All languages mark cognitive distinctions, but different languages mark them in different ways. The conceptual domains of color, space, and time take different representations in diverse lexicons. The sorting of objects, as well as the explanation of actions and events, have been found to vary among cultural and sub-cultural groups (Atran, Medin & Ross, 2005; Bowerman & Levinson, 2001; Gentner & Goldin-Meadow, 2003; Medin & Atran, 2004; Nisbett, 2003).

As pointed out above (2.2.1), cultural differences can leave traces at the neurological level: the logographic orthographies used in Asian cultures recruit neurological resources differently than the alphabetic orthographies common in the West, as evidenced by reading disabilities that pattern differently in the brain depending on which type of orthography an individual has learned to read (Siok, Perfetti, Jin & Tan, 2004; Tan, Laird, Li & Fox, 2005). Language-processing areas of the brain have also been found to be involved in the perceptual discrimination of colors, consequent to the acquisition of a color lexicon, providing some neurological support for the Whorfian claim (Tan, Chan, Kay, Khong, Yip & Luke, 2008).

Despite these findings of variation across cultures, a strong version of the cultural relativity hypothesis has not re-emerged. Quite simply, cross-cultural variation occurs against a background of substantial similarity (Atran, Medin & Ross, 2005; Sperber & Hirschfeld, 2004). Variation among cultures can thus be accounted for without sacrificing a universality assumption.

---

[14] See Caruthers, 2002, for a discussion of the role of language in central, or higher-order, cognition; see also 7.1, this volume.

### 2.2.5 *Cultural Transmission*

We noted above (2.1) that a number of culture-specific factors could have influenced the transmission of textual form and content within the cuneiform scribal tradition. Universals of human cognition and communication would also have influenced the transmission of cuneiform texts within the scribal tradition, since the same factors that govern cultural transmission in modern cultures would have influenced the Mesopotamians.[15] On a naturalistic account, although writing is cultural in origin, and orthographies are cultural representations, they should be able to take a naturalistic explanation.

The survival of ideas in populations across time has been explained as *cultural evolution*, by analogy to biological evolution (Dawkins, 1976, 1982; Boyd & Richerson, 1985; Richerson & Boyd, 2005); or, alternatively, as *cultural transmission*, by analogy to ordinary communication and cognition (Atran, Medin & Ross, 2005; Sperber & Hirschfeld, 2004; Sperber & Claudiere, 2006). Other views draw simultaneously on both evolutionary and communication theory (Tomasello, 1999). The goal of such theories is to account for both the *stability* of cultures (the maintenance of coherent cultural identity across time) and *change* within cultures over time.

On evolutionary accounts, meaningful cultural content has been described as consisting in *memes*, analogous to genes in biological evolution, which undergo transformation in a similar way. That is, stability can be accounted for by replication, and change can be governed by the same sorts of mechanisms that govern mutation or adaptation in biological organisms. Difficulties arise where analogies between biology and culture break down: it has been pointed out that the outputs of an individual's cognitive processes are not usually copies, but rather transformations, of inputs (Sperber & Claudiere, 2006), and thus strict replication doesn't apply.

On cultural transmission accounts, *stability* is maintained by species-level cognitive characteristics, biologically-prepared mental mechanisms that limit variation to readily transmissable psychological forms. Change is accounted for by alterations attributable to ordinary cognitive and communicative processes—memory limitations, for example, or misunderstanding. Cultural transmission is thus partly preservative

[15] See Chapter 1, 1.6, for a comparison of Mesopotamians and moderns.

and partly constructive, and it does not exclude biological explanation. Cultures change, but they also maintain stability across time because they are grounded in core, universal properties of human cognition and communication. The survival of ideas, and the degree and type of change they undergo, is constrained (although not determined) by universal properties of the human mind.

A naturalistic approach, then, seeks to ground cultural explanation in what is known about the biological substrates of human cognition: human mental abilities make culture possible and must, to a meaningful extent, influence its content and organization. Content and form in the cuneiform scribal tradition is no exception. That is, we could expect that the cognitive and linguistic factors that governed the creation of cuneiform series such as MUL.APIN would reflect the same processes of transmission.

## 2.3 *Writing and Conceptual Change*

A cultural invention such as writing clearly can engender conceptual change at the level of orthography and its related linguistic and cognitive discriminations. We see this in both the organization of the lexical lists (2.1.2, above) and in current cognitive-linguistic research (2.2.1, above). It is somewhat more controversial to claim that writing alters conceptual categories that are not directly related to the requirements of using a script.

Logical and scientific concepts have no demonstrable relation to particular orthographies, yet writing has been said to figure in their development.[16] Why should this be the case? Many genres of discourse have appeared in written form since texts first began to appear, including myth, poetry, divination, and religious works, as well as practical handbooks. Why should writing bias thought toward the rational and scientific? If writing engenders conceptual change of a more general sort, toward rationality in general, or toward logical, scientific thought in particular, what sort of evidence could we expect to find in MUL.APIN, and what sort of explanatory model might be able to account for it?

---

16 See Harris, 2009, for a recent comprehensive review of the history of this idea.

We address these issues in turn: first, we review theory and research on writing and cognition; we then advance a model of writing and conceptual change.

### 2.3.1 *Writing and Rationality*

Does writing make us rational? Must one be literate in order to be logical? The idea that literacy and rationality are cultural fellow-travelers permeates the Western intellectual tradition but remains largely unexamined. Harris (2009) refers to this as "scriptism": the assumption that the modern literate mind is civilized, rational, logical, and enlightened, while the primitive, or pre-literate, mind either lacks certain cognitive capacities or manifests them in naïve or unsophisticated ways.[17]

On a naturalistic account of mind, binary distinctions of this nature have no currency. Minds can change, and the functional organization of brains can change, but this can easily be accounted for without positing different *kinds* of minds. It's just ordinary learning. The cumulative cultural effects that led anthropologists and linguists to postulate the distinction in the first place can be accounted for by cultural transmission and cognitive evolution. On our current state of knowledge, there are no grounds for positing a primitive mind that is distinct from the modern mind,[18] yet the assumption lingers on.

### 2.3.2 *The Greeks and the "Great Divide"*

The distinction between ancient and modern minds has been commonly located at a particular cultural-historical juncture: classical Greece. And it can hardly be an accident that a flowering of rational discourse co-occurred historically with the rise of Greek literacy. This observation led to the so-called "literacy hypothesis": writing was a causal factor in the development of logical thought and analytic-reflective modes of discourse in ancient Greece.[19] However, co-occurrence

[17] See, for example, discussions in Boas, 1911; Levi-Strauss, 1962; Goody, 1977; and Harris, 2009, for a current review of the "doctrine of the primitive mind."

[18] See also Renfrew, 1994, for a similar discussion.

[19] See Harris, 2009, for the notion of the "great divide" and an extensive review of the issue; see also Goody & Watt, 1968; Goody, 1977; Olson, 1994, 1996; Harris, 1986:38.

does not constitute a causal argument. That is, while the historical appearance of alphabetic literacy and the rise of scientific, rational, logical thought were contemporaneous, one did not necessarily cause the other. The development of logic and rational argumentation in ancient Greece was realized to a substantial degree within spoken language contexts (Lloyd, 1990) and the impact of alphabetic literacy may have been overstated.

In a strenuous critique of the "literacy hypothesis," Halverson (1992) denies that the medium of writing could have any cognitive effects distinguishable from those of ordinary spoken communication. The effects of literacy, he argues, are due to the *content* of what is written and read, rather than the *fact* that it is written and read:

> The *medium* of communication—which is the issue here—has no *intrinsic* significance in the communication of ideas or the development of logical thought processes.
>
> Halverson, 1992:314; italics in original

The particular relevance of this controversy to our analysis of MUL. APIN is illustrated in a set of claims and counter-claims regarding the relation of the cuneiform lists to hierarchical logical categories.[20]

> Goody takes a new approach by examining lists and their significance (1987:81). While not denying that lists sometimes occur in oral cultures, he finds them a much more distinctive feature of literate societies, appearing with great frequency in the earliest eras of Sumerian and Babylonian writing. Lists . . . thus visualized . . . encouraged the ordering of the items, and activity 'which encourages the activities of historians and the observational sciences, as well as on a more general level, favouring the exploration and definition of classificatory schemas' (p. 108). Lists, then, represent 'formal, cognitive and linguistic operations which this new technology of the intellect [*writing*][21] opened up' (p. 81).
>
> This is an interesting idea, for taxonomy or categorical classification is fundamental to scientific and logical thought. Lists are not uncommon in oral societies (Vansina, 1973:151–4), but do not necessarily have categorical potential.
>
> Halverson, 1992:307

[20] We are grateful to Halverson, 1992, for his clear articulation of this issue, and cite his discussion at length.

[21] Our addition to the original; writing is the technology of the intellect to which Goody refers.

In his criticism of this idea, Halverson (1992) cites examples of lists that appear in oral traditions, such as inventories, genealogies, and itineraries, which have little heuristic value, and in which category structure is either absent or inchoate. He suggests that early cuneiform lists are not unlike the lists of oral cultures (Biebuyck & Mateene, 1971) and that "no new cognitive development" is implied as a consequence of literacy:

> In general, ancient Near Eastern lists or list-like documents show few traces of vertical, hierarchical principles of ordination, of a kind that would make them truly distinctive from the products of oral cultures. [They]...do not seem to reveal any notable differences 'in modes of thought, or reflective capacities, or...cognitive growth' (Goody, 1977: 111). Systematic, hierarchically-ordered classification had to await the efforts of Plato and Aristotle. Plato's 'method of division,' expounded in the *Sophist* and the *Statesman*, seems to be the first major attempt at logical classification. That it was in any way inspired by list-making practices is extremely doubtful.'
>
> Halverson, 1992:308

However, on linguistic criteria alone there can be no doubt that the cuneiform lists differ substantially from "lists" in oral cultures. The typology of lists presented by Civil (1995) has no parallel in cultures that lacked writing. Many of the lists demonstrate a high degree of reflection on the structure of language and its representation in graphic form, and are presented in a logical and systematic manner. Even the encyclopedic entries in the Urra = *hubullu* series, while not co-extensive with the categories of a modern taxonomy, are coherent categories when one takes into account the context within which, and the purposes for which, they were composed: scribal education and reference (cf. Veldhuis, 2004:82). A particularly striking example is the list known as "Proto-Ea," which lists different semantic values for the same written signs, teaching the abstract concept of polyvalence (2004:84).

Thus, while it is fairly certain that knowledge was orally transmitted within the community of cuneiform scribes, there is no possibility that the lists of the archaic cuneiform corpus could have evolved through oral transmission alone. They manifest forms that are uniquely literate.

Halverson's criticism is of particular note, however, because it is representative of the widespread conception that classical Greece represents a historical division between pre-logical and logical discourse;

between pre-scientific beliefs and objective, empirical inquiry; and between the primitive and the modern mind. Harris (2009) points out that this prevalent conception may be due, in no small measure, to the Greeks themselves. The ancient Greeks viewed all non-Greek speaking peoples as barbarians (cf. Havelock, 1982) and as lacking "logos" (Herodotus, cited in Harris, 2009:21). Greek attitudes to their surrounding cultures were embodied in their texts, and the classical works of Greek philosophy and science formed the basis of the Western intellectual canon until modern times. The common conception of ancient Greek culture as the origin of rational thought, of logic, and of empirical science may thus have entered modern conceptions through the Greek texts.

The great divide, in short, is not a divide at all.[22] There is no sound basis for a distinction between a pre-logical and a logical mind, a primitive and a rational mind, or a pre-literate mind and a literate mind. On current cognitive evolutionary criteria, ancient minds were much the same as modern minds. The project of deciding exactly when rationality and science emerged is rendered almost impossible by the difficulty of defining the concepts unambiguously. The search for the origins of science, as Harris (2005) has argued, may be as much a semantic issue as an historical one. That is, what precisely do we mean by "science" or "rationality"?

Another problem for the notion of ancient Greece as a great divide is the increasingly detailed understanding of the broader cultural context of the Ancient Near East and the eastern Mediterranean during the first millennium. Greek thought and science did not emerge as a sudden flash of insight in an Olympus-like splendor of isolation.

The eastern Mediterranean during the first millennium was part of a complex, interactive stream of commercial, political, and cultural exchange that stretched over wide swaths of territory (Gunter, 2009), a spread of ideas so extensive as to be comparable to the modern effects of globalization (Pongratz-Leisten, 2010). This cultural "cross-pollination" can be traced through the diffusion of art and visual motifs, religious ideas, written texts, different languages and the diverse orthographies adapted to represent them (Rubio, 2007). The Greeks, along with other cultural groups in the region, would have been swept

[22] Harris' (2009) book-length discussion of this issue is both concise and comprehensive.

up into the mix of commerce, competition, and creativity. The spread and survival of cultural content in those complex times would have been governed by much the same factors as in the modern period. It is difficult to maintain that any one culture was unaffected by the others. Cultural boundaries were porous; languages and orthographies, along with ideas and other cultural contents, were widely exchanged. Any consequences of literacy, then, were unlikely to have been realized only in a single culture, or limited to a single orthography.

### 2.3.3 *Moderns, Media, and Materialism*

The "literacy hypothesis" was not an isolated claim. It bears a family resemblance to a more general claim about the impact of technologies of communication on human cognition. Prior to Goody's articulation of the hypothesis, the idea appeared in two widely disparate intellectual contexts in the early-to-mid twentieth century.

The Toronto School[23] dates from the work of Innis (1951), who came at the issue from the field of economics, and McLuhan (1962; 1964) who approached it from the perspective of literary criticism. The main tenet of this school of thought is that the medium of communication—writing, print, or modern electronic media—has effects that are distinct from the content of the message they are used to convey. The most extreme statement of this view was found in McLuhan's famous dictum, "the medium is the message." While discounted by academic contemporaries, their ideas influenced, directly or indirectly, subsequent scholarly work and helped to define the field of communications as a modern academic discipline.

A second, independent set of circumstances yielded the related hypothesis that the technology of literacy was central to engineering cultural and cognitive change. A group of psychologists caught up in the ideas of dialectical materialism in the post-revolutionary Soviet Union were also led to contemplate, and empirically investigate, the effects of literacy on thought. In the early part of the twentieth century, the Soviet Union began a programmatic drive to raise the level of

---

[23] See Watson & Blondheim, 2007, for a discussion of the Toronto School and the influence of Innis and McLuhan.

popular literacy among its citizens. Alexander Luria (1976),[24] a Soviet psychologist, mounted a research expedition to the remote republic of Uzbekistan to explore the cognitive consequences of literacy. His empirical investigations compared recently-literate Uzbek adults with their unschooled agrarian fellows on a range of tasks, including syllogistic reasoning and hierarchical category organization.

The participants in his research were asked to sort triads of pictures of conceptually-related objects, such as a log of wood, a hammer, and a saw. When asked which two pictures "went together," the literate individuals would choose hammer and saw, since both are tools. The non-literates, in contrast, tended to choose the saw and the log of wood, since a saw would be used to fell the log. These different sorts of choices were argued to illustrate a bias toward hierarchical logical classification among the newly literate, in contrast to a more functionally-biased classification strategy among their unschooled fellows.

In addition, newly literate Uzbeks were able to reason from verbally expressed premises, along the following lines:

> In the far North, where there is snow, all bears are white.
> Novaya Zemlya is in the far North, and there is always snow there.
> What color are the bears there?
>
> Luria, 1976:108–9

Literates reasoned according to the expressed premises: the bears would be white. Non-literates responded along the lines of the following: "I don't know... There are different sorts of bears."

Non-literates, in other words, tended to rely on direct apprehension and experience rather than reasoning from premises. Adult literates, even after only a few years of schooling, reasoned from linguistically expressed premises. Luria concluded that his results showed an increase in logical reasoning ability and hierarchical category organization by the literate group. Literacy, he thought, was the causal factor that differentiated the cognitive preferences between the literate and non-literate groups. This conclusion may have been strongly influenced by the Zeitgeist in which he and fellow psychologists Vygotsky and Leontev were then steeped (Lloyd & Fernyhough, 1999).

[24] Luria's research was noted briefly in the *American Journal of Genetic Psychology* in 1934, but internal political censorship prevented a full reporting of his results until many years thereafter, in 1976; see discussion in Lloyd & Fernyhough, 1998.

Luria's binary distinction of "pre-literate" and "literate" suggests, to some degree, his own cultural bias, or "scriptism" (Harris, 2009). The unschooled subjects of his experiments had their own logic, but it did not mirror his conception of the word-based logic acquired in formal educational contexts. On Harris's (2009:170–1) analysis, rationality takes varying forms in various cultural contexts. Broadly defined, it hinges on the sign-making capacity that human beings exercise in their communications with others every day of their lives, and it underpins social organization in both literate and preliterate cultures. Rationality, then, has its own pragmatics, and is relative to contexts of activity. This is supported by current empirical research, which shows that diverse cultural experience of any sort can influence categorization, even when individuals share the same immediate contexts of activity (Medin, Ross, Atran, Cox, Coley, Proffit, & Blok, 2006).

### 2.3.4 *Pragmatics and the Uses of Writing*

An attempt to replicate Luria's findings did not succeed. A large and well-known empirical study was conducted some years later among the Vai in Liberia, by an American research team. Scribner and Cole (1981) distinguished three degrees of literacy within the Vai culture. One group of the Vai used writing for very limited purposes, such as composing personal letters. Two other groups were schooled, one on the Western text-analytic model and the second on a more traditional model that emphasized the memorization of texts. Scribner and Cole administered a battery of tests based on those used by Luria, and found that only the group schooled on the Western text-analytic model performed in a manner similar to the literates in Luria's study. Their conclusion was that *schooling*, rather than literacy, was implicated in cognitive change. An increase in syllogistic reasoning and hierarchical category organization, they argued, was a consequence of the *cultural practices* of which literacy is a part, rather than literacy alone.

These results were cited in Halverson's (1992) conclusion to his critique of the literacy hypothesis: "the consequences of literacy depend entirely on *the uses to which literacy is put*" (1992:314). The only other observation about possible effects of literacy that survives Halverson's critique is that of *permanence*: it is the permanence of written records that is responsible for the role that writing plays in a "cumulative intellectual tradition" (1992:303). He takes pains to emphasize that neither claim, whether about *use* or *permanence*, entails an assumption of *cognitive effects* of the medium of writing. But is this so?

### 2.3.5 *Permanence, Memory, and the Archival Uses of Texts*

The cuneiform corpus attests to the ancient origins of the archival function of writing and its importance in cultural transmission. Assurbanipal's library may be the best known, but other sites, such as Ebla, have yielded library-like collections of greater antiquity. The existence of finely inscribed tablets dating to the Old Babylonian period (Black & Tait, 1995) and extra-curricular lists executed on large tablets in beautiful writing (Veldhuis, 2004:89) suggest that cuneiform tablets were objects of value that were collected in private homes as a sign of status, and not simply products of quotidian labors at the scribal academy.

Permanence of representation is the defining property of the medium of writing and is at the core of our common-sense understanding of its functions: record-keeping, transcription, and the storage of manuscripts and books that cumulatively form the archives of a civilization. Written archives allow an accumulation of cultural contents to an exponentially greater degree than that allowed by spoken language and significantly enhance cultural transmission. This enhancement is due to a concomitant release from the limitations of human memory.

Writing as an aid to *memory* is attested from the earliest uses of writing (Harris, 2009:121), and as noted above, writing was invented in response to the needs of administration and accounting (Nissen, Damerow & Englund, 1993). The fallibility of human memory, and the need for an objective, authoritative, and permanent account of transactions that would be recognized by all parties involved, may have been the single most important reason for the invention of writing. Writing as an aid to memory is recognized equally by those who advanced the literacy hypothesis and by its critics (cf. Halverson, 1992), and the effects are twofold: long-term storage, and immediate reduction of processing load.

Cognitive models of human memory are based on a general division between long-term memory and short-term, or working, memory. An individual's long-term memory is the sum total of remembered experiences, while working memory consists in the information that can be kept active while a particular problem is being solved. The archival property of writing enhances both. The stored, cumulative information of a culture in an archive is analogous to a shared, accessible "long-term memory." But as Halverson (1992) notes, it is not immediately evident how the mode of storage—the *medium* of writing—has cognitive effects.

The distinct properties of short-term memory clarify this: the use of writing, whether in the form of a book, manuscript, or notational device, renders the recorded information accessible "on-line," to use a computational analogy. It can be used in problem-solving without taxing the limited resources of short term, working memory.[25] Writing reduces the *load* on working memory, and increases the amount of information that can immediately be brought to bear on the task at hand. Quantitative change of this sort can lead to qualitative change.[26] Is this reduction of load on short-term memory a "cognitive effect" of writing?

Halverson (1992) argues that the memory enhancement afforded by archives and the permanence of writing are unrelated to cognitive effects. This might appear to be so on first glance. Archives, such as libraries, are free-standing, physically distinct, and apparently autonomous entities. However, they are inert in the absence of minds prepared to read, interpret, and otherwise appreciate their contents. The classical archive did not cease to exist after the Vandals sacked Rome. The so-called Dark Ages were not dark because of the *absence* of archival knowledge, but rather because the archives that existed were not being *read and interpreted.* An archive that remains inert and unused can have no effects—cognitive, cultural, or otherwise. There is no substrate upon which the effects of archives can be realized other than human cognition.

Cultural and social change may be the end or cumulative result of the archival storage and transmission of knowledge, but it isn't clear how either of these could be realized in ways that are not directly consequent to human cognitive activity. Any effects that are attributed to the permanence of written records are thus necessarily consequent to the *medium* of writing.

Archives are uniquely enabled by the permanence of written representations. But the effects of archives can only be realized through their use and interpretation by the human mind. It follows that an

---

[25] Short-term memory is limited to around seven chunks of information, plus or minus two (Miller, 1956), see also Baddely, 2007.

[26] See Chapter 7, 7.1, for working memory in dual-process models of cognition.

argument for *archival* effects necessarily entails an argument for *cognitive* effects, and that those cognitive effects are necessarily contingent on *the medium of writing*. Archives, in short, have no currency in the absence of minds prepared to interpret them. Archival effects are thus cognitive by definition.

## 2.4 *A Model of Writing and Conceptual Change*

The role of writing in cultural transmission, and its importance in understanding the nature of texts such as MUL.APIN, can be clarified by developing a model of writing and conceptual change. Such a model would necessarily draw on theories of cultural transmission that appeal to both cognition and spoken language. We begin by considering the role of spoken communication in cultural transmission, and then consider the distinct properties of writing: can we expect writing to alter the process of cultural transmission in predictable ways? Is there any reason to expect that writing alters conceptual content in a manner different from ordinary spoken communication? We show how the *permanence* of written representations and their *use* are interdependent. This interdependence suggests the *mechanism* by which writing engenders cognitive effects: it *recalibrates* inferential environments.

### 2.4.1 *Writing and Cultural Transmission*

The archival function of writing is, itself, not in dispute. It follows directly from the permanence of written representations, and the accumulation of knowledge and cultural contents that permanence allows. But beyond this preservation, and the effects on memory that it clearly has, does writing have an influence on the *processes* of cultural transmission that can be distinguished from that of ordinary spoken language?

The role of spoken language in cultural transmission has been widely acknowledged.[27] Language enables us to take advantage of the minds of others, and thus allows access to the cumulative knowledge of a culture and the transmission of knowledge to the next generation. Writing differs from spoken language in that it is not universal—it appears only in some cultures—and has different representational properties.

[27] Dennett, 1994; Tomasello, 1999; see discussion in Chapter 2, 2.2.

Principal among these is the permanence of written representations, as we have seen. This contrasts with the evanescence of spoken language. Permanence allows the creation of cumulative repositories of cultural knowledge, or archives. In contrast to spoken language, then, writing allows access to the *contents of minds from prior generations*, and to the minds of individuals the reader has never met. Writing could thus be expected to enhance, even to accelerate, the processes of cultural transmission. Understanding how it achieves this is illuminated by a comparison with the role of spoken language.

### 2.4.2 *Writing as Communication*

Harris (2009) argues that *writing is a form of communication sui generis*. Writing systems were initially invented to convey information,[28] and an examination of the earliest written symbols clearly shows that the relation of written signs to speech was not a given, but was discovered over the course of centuries. The cuneiform list corpus, from the archaic Uruk period through the Old Babylonian period, attests to this process of discovery. The orthographic representations of linguistic categories became organizing principles in many of the lists (2.3, above), but representing linguistic categories was not the impetus for the invention of written signs. Signs had to be adapted and re-adapted in order to approximate usable representations of the different spoken languages of the second and first millennia (cf. Rubio, 2007).

If we assume that writing is communication *sui generis*, what related assumptions necessarily follow? On cognitive accounts of communication, the ideas advanced by Grice (1968; 1975)[29] have been influential. Grice shifted the focus of philosophical debate on the nature of meaning from *linguistic* to *mental* representation. He differentiated between the linguistic meaning of words, or "sentence meaning," and what a speaker means by the use of those words: "speaker's meaning."[30]

A speaker, on Grice's account, intends to produce an effect in the hearer by means of the *recognition of his intention*. He uses words to realize

---

[28] See Harris, 1986, 2000, 2009; and Olson, 1994, 1996, 2001 for related discussion.

[29] The distinction between speaker's meaning and utterance meaning is found in Grice, 1968; a discussion of conversational implicatures and maxims is found in Grice, 1975; Grice, 1989, contains a collection of prior work.

[30] The pragmatic tradition, notably the work of Wittgenstein, Austin, and Searle, is of direct relevance to this insight.

his intentions, and the hearer exploits the conventional meaning of those words when interpreting the speaker's intention. But the process of utterance interpretation also involves non-linguistic inference. In ordinary spoken conversation, inference is based on a shared physical context, existing knowledge, prior utterances, and some understanding of conversational conventions. Post-Gricean models of communication[31] similarly assume a mix of linguistic decoding and inferential processes, but more heavily emphasize inferential processes. There remains some controversy, such as over the extent to which inferential processes are either conscious and reflective, or intuitive and non-reflective.[32] For the purposes of our present discussion, this finer level of distinction is less important than the more basic distinction between *language-based decoding* and *non-language-based inference*.

On inferential models, the hearer, or receiver of the spoken message, will draw on all available sources of information in the process of utterance understanding. The linguistic form of a message is clearly a rich, or privileged, source of information, but the *context* of an utterance, the occasion, time, and person speaking are also relevant to utterance interpretation, and are processed simultaneously. Context is not restricted to the physical environment that two interlocutors may share. Rather, it includes the *entire set of assumptions* that a hearer may bring to bear on the process of utterance interpretation, including the *mental representations* derived from the perception of the physical environment, interpretations of prior linguistic utterances, prior knowledge, beliefs, event representations, memories, and the like.

Context, on inferential models such as Relevance Theory (Sperber & Wilson, 1995), is thus *cognitive*. Utterance interpretation crucially depends on assumptions about intentions, or, *what a speaker has in mind*. The notion of context as cognitive has particular importance for understanding the role of writing in communication, because the effects of writing have often been linked to "decontextualization": the property of writing that allows message interpretation to take place in the absence of a shared context.

---

[31] See Breheny, 2002; Carston, 2002; Sperber & Wilson, 1995, 2002, for cognitive pragmatic accounts of inferential communication.

[32] We don't address these controversies, as they are not relevant to the present concerns of this chapter; nor do we attempt to address the complex issue of how these models interface with cognitive architecture, but see Chapter 7 for further discussion.

In ordinary communication, speaker and listener are both present. Thus, in addition to the linguistically-encoded message, or what a speaker actually *says*, a listener has access to un-encoded information: his knowledge and assumptions about the speaker, the physical environment that they both share, and so on. This is not true of written communication. Reader and writer *need not*, and in practice, usually *are not*, both present during text understanding, and hence, the interpretation of writing is described as decontextualized.

But if context is cognitive, as on current inferential models, context can never be absent, and interpretation thus can never be "decontextualized." The absence of a shared physical context between writer and reader, rather, necessarily entails *a reduction in the non-linguistic sources of inference* that can be brought to bear on interpretation. This reduction of non-linguistic sources of inference co-occurs with the continuing availability of the written representation. Thus, the aspects of the message that can be written—or, encoded in language—assume a greater importance in extended inferential processes. In other words, the effects of the permanence of writing are most evident in extended inference, or re-interpretation.

When a message is ambiguous, abstract, or otherwise problematic in interpretation, the written form allows *re-interpretation* to a degree not afforded by spoken utterances. This re-interpretation would be along the lines of extended, or reflective inferences that occur in ordinary spoken communication when spontaneous inference fails to yield a satisfactory interpretation (cf. Sperber & Wilson, 2002). The advantage of a permanent written representation is thus most obvious at the level of conscious chains of inference, of the type involved in logical reasoning or calculation.

### 2.4.3 *Writing Recalibrates Inferential Environments*

By preserving linguistic form, writing alters the degree and kind of interpretation that can be carried out. On an inferential account of communication, this would invest a written representation with a *higher information value* than the same linguistic form might have in a spoken exchange. A spoken message is evanescent: a hearer may repeatedly try to puzzle out what the speaker intended, but the actual words that were spoken only remain available if they are actively rehearsed in short-term memory. In contrast, by isolating and fixing the encodable linguistic elements of a communication event, writing weights the linguistic code more heavily when non-linguistic sources of inference are absent. In this way, *writing recalibrates the inferential environment in which communication takes place.*

The effects of recalibration are relative to the inferential demands of the message. Simple messages, whether written or spoken, rarely require re-interpretation. If an intended meaning is easily understood, re-interpretation would not be expected to occur. When writing takes the form of personal letters between mutual acquaintances, notes to oneself, or records of simple commercial transactions such as the sale of wheat or salt, it is easily understood. The medium of writing does not advantage interpretation in any obvious way other than as an aid to memory. Scientific or procedural texts, on the other hand, would present a higher interpretive demand. Complex, ambiguous, or abstract expressions usually require re-interpretation. In cases such as this, a written representation provides a distinct advantage.

This accounts for the findings of the Scribner and Cole (1981) study. The simple, personal uses of writing, such as letter-writing and uses of text for memorization resulted in no measurable cognitive changes. Changes were found only when texts were analyzed in formal educational contexts, which involved higher interpretive demands. Under these conditions, writing could be expected to have measurable cognitive effects. This is indeed what was found.

Scribner and Cole attributed the effects to schooling and cultural practice, which may reflect the difficulty of isolating and empirically quantifying cognitive factors in the multi-faceted educational environments they studied. Field research typically cannot achieve the same exactitude as lab research. But their findings cannot be taken as constituting conclusive evidence against the cognitive effects of the medium of writing, or of literacy *qua* literacy. Such effects could easily have been realized within the context of formal schooling.

If the effects of writing were due to "decontextualization" alone, any written representation would have more or less the same effect. A recalibration account, in contrast, predicts that simple uses of writing will have no cognitive effects. Permanent written representations present no processing advantage when inferential demands are low, as in the use of personal letters or rote text memorization. The advantage of a permanent representation would manifest only with increased inferential demand. The more analysis and re-interpretation required, the greater the effects that the medium of writing could be expected to have. *The effects of writing as a medium of communication, then, are dependent on the uses to which writing is put. Recalibration is relative to inferential demand.*

### 2.4.4 *Writing and Rationality*

Why, and how, should writing influence rationality? An inferential account suggests the mechanism whereby such cognitive effects may be realized.

First, writing does not *necessarily* lead to cognitive effects. It rather *allows* cognitive effects by recalibrating the inferential environment within which interpretation takes place. Recalibration is relative to inferential demand. That is, the effects of *permanence* and *use* are interdependent. Conceptual change may result from the extended inference that writing allows, but extended inference does not always take place. Only the interpretation of complex, ambiguous, or abstract messages is significantly advantaged by writing, because in these cases, the permanent representation of linguistic form provided by writing remains available for re-interpretation, reflection, and analysis, while non-linguistic sources of information (such as the shared physical context and social interaction of spoken conversation) are simultaneously reduced or absent.

Texts that present high interpretive demands are likely to require extended re-interpretation. When this occurs, the permanent representation of linguistic form in writing assumes greater weight than the ancillary properties of speech, which cannot be so represented, in the interpretation process. This weighting of linguistic form entails a bias toward *logical form*.

Logical form (LF), loosely defined, is a mental representation of the meaning of a sentence; or a semantic rendering of its structure (see Fox, 2003, for a comprehensive account of logical form and the difficulties associated with defining it unequivocally). LF underwrites the inferential properties of a sentence. The syllogistic reasoning abilities that Luria (2.3.3, above) attributed to literacy illustrate this sort of inference from LF. That is, to the extent that LF is recoverable from linguistic form, preserving linguistic expressions in writing renders LF more accessible for extended inferential processing. By isolating and permanently representing linguistic form, and thereby rendering LF more accessible, writing recalibrates inferential processes toward LF. In this way, over time, writing could be expected to bias the mental representations of literate individuals toward representational content that can be underwritten by LF. The inferential model, then, suggests how and why writing might lead to the enhanced development of logic, or rationality.

Man, by his own description "the rational animal," may thus be made more so as a consequence of his own invention—writing. Communication, as noted above (2.2.3) is common among other species, and human communication shares features with the animal variety, notably the social and interactive aspects. The property that distinguishes human language as unique is the generativity, or creativity, enabled by syntax, the combinatorial rules of language (or recursion, cf. Hauser, Chomsky, & Fitch, 2002) that allow an infinite number of new sentences to be generated and understood. The aspects of communication shared with other species are exactly those *not* represented in writing. It seems curious that a cultural invention should isolate biological uniqueness, but to the extent that writing isolates the logical form of language, this is what occurs.

It should be emphasized that "logical form" in the Chomskyan, linguistic sense is not commensurate with scientific logic, or other senses of the word "logic" (Kneale and Kneale, 1984). However, the relation of language to logic, and the pivotal role of language in the development of the formal disciplines of science and mathematics, has filled volumes. The predicate calculus was an attempt to 'bring language to heel,' so to speak, in the service of the exact disciplines for which it was used, often in an unexamined way.

> the logic of science is nothing other than the logical syntax of the language of science.
>
> Carnap, 1937:xiii; in Harris, 2005:167

As Harris (2005:168–9) describes it, the heart of Carnap's enterprise was "ultimately to construct a language of science which is free from the confusions, imprecisions and deceptions of 'ordinary language.'" Harris contends that Carnap and the Vienna Circle ultimately failed in this program, but failure or success may be less important than the attempt itself. That is, it is the logical positivist program in which they were engaged reflects a uniquely literate sensibility.

Content that can take explicit expression in language can, in theory at least, take a logical form. It may do so imperfectly, but its expression in writing allows LF to be recovered and scrutinized. A spoken utterance, in contrast, is evanescent. Conversation is more directly apprehended, and its properties are more elusive.[33] Writing, then, necessarily biases the literate mind toward logical form by isolating linguistic form. This occurs whether or not writing perfectly represents language (it doesn't) and whether or not language is strictly "logical", in the sense that Carnap (1937) was after.

Writing, then, does not *necessarily* lead to cognitive effects, but rather creates the conditions that allow cognitive effects to occur. Simple uses of writing, such as letters or notes, are unlikely to have any effects. Complex, vague, or abstract texts, could be expected to lead to cognitive effects, because writing *recalibrates* the inferential environments in which complex texts are understood. Texts with high inferential demands bias the literate mind toward the logical, by isolating and permanently representing linguistic form and rendering LF more accessible for extended inference.

The astronomer-scribes of Mesopotamia who compiled the MUL.APIN treatise were part of a scribal tradition that had developed a highly articulated sense of written language, as evidenced, for example, in the lexical lists (Civil, 1995). Both written form and thematic content, in a similar manner, are likely to have shaped a range of text traditions in the cuneiform corpus, as they were read and copied through generations and over centuries. The form that the canonical MUL.APIN treatise ultimately came to assume is thus likely to reflect generations of cultural transmission. To the extent that cultural transmission is limited to readily transmissible psychological forms, the text of the treatise should reflect the thought processes and conceptual organization of the astronomer scribes who composed it.

## 2.5 *Conclusion: Summary of Pre-Analytic Assumptions*

Our analysis of MUL.APIN is based on the following assumptions:

1. *Canonical MUL.APIN reflects, in part at least, the content and linguistic forms of the cuneiform astronomical tradition of which it is a part.* MUL.APIN developed *cumulatively*, beginning with simple textual forms such as the star lists, to which were added more complex and sophisticated

[33] Grice (1975) observed that conversation has its own logic. The logic of conversation manifests in conventions, or maxims, that bear on relevance to a topic and quantity of information. An utterance that supplies too much information, or too little, or that is not immediately relevant to the topic of conversation, is not effective.

textual components over time. The treatise was compiled within a scribal tradition that spanned generations and centuries. The canonical version represents an omnibus collection of astronomical texts, and includes content—in some cases, complete sections of text—that were preserved and transmitted within the framework of this scribal tradition, although surviving written precursors to MUL.APIN are for the most part not extant.[34] This process culminated with the writing and "publication" of the canonical series.

We set forth at length, in our discussion of the Mesopotamian text tradition,[35] our reasons for assuming that canonical MUL.APIN is based on previous written material. Parallel examples can be found in the transmission of the Gilgamesh Epic, where parallel sections and exact, or near exact, wording were preserved at intervals of centuries, although the intermediate stages of transmission are not represented by extant surviving manuscripts (George, 2003; Tigay, 1982).

Similar arguments can be made for other text traditions in Mesopotamia. Perhaps most relevant for the case of MUL.APIN is the transmission of the Astrolabe tradition and its constituent parts from the middle second millennium down to the Hellenistic Period, as documented in written sources (Horowitz, 2007). MUL.APIN seems unlikely to have evolved only in oral form, and it seems equally unlikely that it could have been written in a single series of sittings, at a single point in time, by a single author or group of authors, as is generally accepted for the Babylonian national creation epic Enuma Elish (see Lambert 1960).

2. *Canonical MUL.APIN reflects a process of cultural transmission, and should accord with a general model of cultural transmission.*[36] Stability and change in cultural transmission can be accounted for by human cognition and communication. Stability is accounted for by invariant core cognitive phenomena: cultural contents that "survive," or persist, across generations are those elements that are compatible with core cognition.

These core cognitive elements form the basis of our analysis of MUL.APIN. Change, or variability in transmission, can be accounted for by a range of cognitive and communicative influences, including perceptual factors, the nature and limitations on representations in

[34] An exception may be found in George, 1991:303–306.
[35] See Chapter 1 and section 2.1, this chapter.
[36] See this chapter, 2.2.5.

short- and long-term memory, environmental factors, and ordinary spoken communication, which is a primary means of transmission in human cultures. Variation, on this account, is similarly limited to readily transmissible psychological forms (cf. Atran, Medin & Ross, 2005).

3. *Writing is a form of communication sui generis, with properties (permanence), uses (archival), and effects that are distinct from those of spoken communication.* A Gricean account of inferential communication suggests how writing has its effects. Writing *recalibrates* the inferential environment in which communication, and thus cultural transmission, takes place. Non-linguistic sources of inference are reduced and written representations are weighted more heavily in inferential processes. Recalibration is relative to use: the interpretation of abstract, complex, or ambiguous expressions is enhanced by permanent written representations, while the interpretation of simple expressions is unaffected.

4. *Writing biases mental representation toward logical form (LF).* By isolating and permanently representing linguistic form, writing weights it more heavily than the ancillary properties of spoken language in interpretation. To the extent that LF is recoverable from linguistic form, writing renders LF more accessible. Writing may, in this way, bias the outcome of extended inferential processes. Over time, writing could be expected to bias the mental representations of literate individuals toward representational content of the sort that is underwritten by LF. This, in turn, may lead to the enhanced development of logic, or rationality.

# CHAPTER THREE

# TERMS OF ANALYSIS

The terms of analysis that we apply to MUL.APIN were derived in large part from the text itself. They are the result of considering the form and content of the treatise in light of current theory and empirical research on language and thought. A multitude of analyses would be possible, in theory, but we restrict ours to the basic categories of language and thought that seem necessary to understand the text.

No analysis can be entirely theory-neutral, but we have attempted to keep the terms detailed below as neutral as possible, in order to open up the text to additional analysis and scrutiny. An ancient astronomical text such as MUL.APIN is relevant to a range of scholarly disciplines, but a naturalistic approach makes possible a degree of neutrality. That is, the terms of analysis we apply should, in principle, be amenable to other approaches and interpretations.

## 3.1 *The Language of Space and Time*

Any coherent description of astronomical phenomena must rely on grammaticalized notions of space and time: those aspects of location, motion, and temporality that can be expressed in language. The perceptual and conceptual distinctions inherent in human cognition create enduring distinctions in language, as is evident from a comparative examination of historical perspectives on space and time, and current psycholinguistic research.

Since the Enlightenment, space and time have been described with a vocabulary that has remained largely unchanged. Newton (1686/7; see Disalle, 2002) argued that space provides an *absolute* frame of reference within which objects move, a view he shared with his predecessor, Galileo. Leibniz (Jolley, 1994), in contrast, claimed that space exists only as a *relation* between objects, with no existence apart from the existence of those objects. The notion that space and time are elements of a systematic, unified framework that we use to structure our experience is found in Kant (1781/1990). In our own day, dating from Einstein's seminal work (Miller, 1997), it has become generally accepted that space and time are continuous.

The content of ordinary language and thought is surprisingly congruous with the principles of physics and philosophy (cf. Dascal, 1987). The same distinctions between absolute and relative views of space can be seen to define the systems of spatial reference that are used across virtually all modern languages.

### 3.1.1 *The Language of Space*

The most basic psycholinguistic distinction in the domain of space is between *figure*, or the object of interest, and *ground*, or the background in relation to which the figure is to be located, perceived or described (Miller & Johnson-Laird, 1976; Talmy, 1983). An analysis of modern languages finds that figure and ground are located in relation to frames of reference, or *coordinating systems*, that fall under three types: *intrinsic, relative*, or *absolute* (Levinson, 2003; Majid, Bowerman, Kita, Haun & Levinson, 2004).

Levinson (2003) points out that some languages employ only one or two frames of reference, while others employ all three coordinating systems. The systems encoded in the language are related to differences in performance of the speakers of those languages on cognitive-spatial tasks. There is some argument about whether the causal factor underlying the difference in performance is language or culture (Li & Gleitman, 2002), but in practice, it is often difficult to isolate the effects of one from the other. On either view, it should be noted that the variation of frames of reference between languages, or cultures, is constrained. That is, no known language has displayed a fourth or fifth type of coordinate system. In this sense, the variability that Levinson (2003) observes can be viewed as relative to substantial underlying similarity, or universality, for reasons discussed in Chapter 2.

The terms used in our analysis of spatial expressions in MUL. APIN are drawn from Levinson's (2003) comparative psycholinguistic approach, which are briefly described in the following section.

### 3.1.2 *Coordinating Systems or Frames of Reference*

*Intrinsic Coordinating Systems*
Intrinsic frames of reference are the simplest. They express binary relations between named aspects of figure or ground:

*beside the car*

Interpretations of these binary expressions are usually based on the canonical orientations of the objects mentioned. Thus,

*the front of the TV*

would refer to the part of the TV on which the screen appears, or "front" with respect to its canonical function. With respect to motion, canonical motion of the figure or ground governs interpretation. Thus,

*the truck moved forward*

would be interpreted according to the truck's canonical trajectory, or how trucks usually move.

*Relative Coordinating Systems*
Relative frames of reference express a ternary or three-way relation. Figure and ground are described with respect to a third point, or perspective, that can be termed a "deictic center" (3.2, below). The deictic center could be a speaker or writer, a listener or putative reader. Thus,

*the cat is behind the tree*

is interpreted with respect to a point of view in which the tree is between the deictic center and the cat. Trees have no intrinsic front or back. A ternary or three-way relation that invokes a deictic center thus determines how we understand *behind* in this example. *Left, right, in front of, back, behind* are lexicalized spatial coordinates that are usually interpreted relative to a deictic center, a speaker or listener.

The advantage of the relative coordinating system is that by specifying a third element, a fixed point of view, a wider range of inferences about locations and movements of figures with respect to ground can be drawn. However, in the broader sense, both intrinsic and relative coordinating systems can be regarded as relative, and both are more clearly contrasted with absolute frames of reference (below).

In our analysis of MUL.APIN, we categorize expressions according to both intrinsic and relative coordinating systems, but in practice, we found that the grounds for distinction between the two systems were not always clear cut. In doubtful cases, we opted for ascription to the relative coordinating system.

#### *Absolute Frames of Reference*

Absolute frames of reference use fixed bearings, objective, or absolute coordinates, such as cardinal directions: *north, south, east* and *west*. In the absence of strict bearings determined by points on a compass, some languages approach absolute coordinating systems by the use of modifying terms. Reference to the direction of the sun's rising or setting, the position of which changes with the seasons, thus isn't strictly a reference to cardinal *east* or *west*, but can function as such when paired with additional information, such as winds or stellar phenomena (Levinson, 1996:371).

Absolute coordinate systems allow the most sophisticated and precise expressions of location and movement, because they are not dependent on a particular viewpoint. Cardinal directions are abstract notions. They are not fixed geographical places, but bearings used in reckoning location, a conceptual grid superimposed on an environment. Absolute coordinate systems thus allow a wider range of inferences than either intrinsic or relative systems. Motion and object alignments can be expressed more easily, without dependence on reference to specific locations or places.

### 3.1.3 *The Language of Time*

#### *The Vocabulary of Space-Time*

Time is universally experienced as continuous and unidirectional. All languages mark temporally bounded events:

> *The meeting was yesterday between two and four.*

and the sequential order of events:

> *He took out the garbage before leaving for work.*

In addition, the dominant modern conception that space and time are two aspects of a unified whole is not simply a theoretical view held by physicists. The distinctions marked in ordinary language illustrate how basic, even universal, is this assumption. Boroditsky (2001) illustrates that in diverse languages, words from the domain of space are frequently used to express *temporal* concepts:

*I'm looking forward to seeing you.*
*The idea is ahead of its time.*
*He's falling behind in his work.*

Similarly, spatial "front—back" metaphors can convey temporal sequence:

*Good times are ahead of us, hardships are behind us.*
*Push the deadline back.*
*Move the meeting forward.*

This space-time mapping is not limited to English. Mandarin uses *front-back* metaphors to describe time (*qián* or front, *hòu* or back). In addition, Mandarin uses *up-down* metaphors to express sequence:

| | | |
|---|---|---|
| *shàng* | *up* | used to express an earlier event |
| *xià* | *down* | used to express a later event |

Thus time, here, flows downwards in Chinese, from above to below.

In Akkadian, the same types of prepositional phrases are used in much the same way as in English although there are some small differences, as will be evident in our analysis of the text. For example, in Akkadian, "above" and "below," which are normally used to indicate "up" and "down" within vertical frames of reference, can indicate "in front of" and "behind" in horizontal frames of reference (Shaffer, 1981). In much the same way, "in front of" can indicate "above/before/aforementioned" in the case of cuneiform text written on tablets, i.e. inscribed before in time, inscribed earlier on the tablet, and when held in the hand, higher up on the tablet (see Horowitz, 1991:77).

### *Space-Time Relation*

Spatial metaphors can provide relational structure in the abstract conceptual domain of time. They are not simply linguistic metaphors, but have cognitive effects. Spatial priming affects the processing time of temporal expressions, and it does so differently depending on metaphors used in the language.

Boroditsky (2001) presented English and Mandarin-speaking subjects with priming sentences that expressed either horizontal spatial relations, for example:

*The black worm is ahead of the white worm*

or for vertical spatial relations, the priming sentence:

*The black dot is above the white dot*

These presentations of priming sentences were paired with visual drawings of the objects named (worms or dots) in two alternate spatial arrangements (horizontal or vertical). In the testing phase of the experiment, after hearing the priming sentence, these same subjects were asked either one of two types of true-false questions. Half of the questions used space-time metaphors:

*March comes before April*

while the other half were purely temporal:

*March comes earlier than April*

If space-time metaphors have cognitive effects, we could expect to find a difference between Mandarin speakers and English speakers, since English does not use the up-down metaphor to describe sequence. That is, if the *up-down* metaphor in Mandarin created a vertical bias in the thought of Mandarin speakers, it should be revealed in different response times to the different types of true-false questions.

The results of this experiment bore out the hypothesis. English and Mandarin speakers were both affected by the priming stimuli, but the responses patterned differently in concert with the different linguistic metaphors. On spatial-temporal *before-after* target questions, used in both languages, both groups of subjects were quicker to respond after horizontal primes. However, on the purely temporal target questions, only the Mandarin speakers responded more quickly after the vertical prime.

Because of the philosophical, scientific, and psycholinguistic continuity between the domains of space and time, we treat the language of space and time together in our analysis of the language of MUL.APIN. We also extrapolate from Levinson's (2003) three frames of reference, or coordinate systems, in order to characterize language that marks distinctions in the conceptual domain of time.

*Frames of Reference, Time, and Space in MUL.APIN*

With respect to our analysis of MUL.APIN, we note that the ancient Mesopotamians conceptualized the four wind directions as ninety degree arcs rather than fixed points. Thus for the ancient compilers of MUL.APIN, *north* referred to a 90° arc ranging from the northeast to northwest, rather than to a fixed co-ordinate point on a compass. Here, then, an absolute frame of reference, in Levinson's (1996) terminology refers to a fixed *range* rather than a fixed point, but can still be regarded as absolute in the sense of being objectively fixed.

Likewise, the seasons in MUL.APIN do not begin at the solstices and equinoxes as do our summer, fall, winter, and spring. Instead, MUL.APIN divides the year into four distinct quarters, each with its own characteristic weather pattern, and corresponding position of the sun in the sky. These begin midway between the solstices and equinoxes. For example, the Mesopotamian hot season (what we might call summer) lasts from May to August, and the cold season lasts from November to February.

## 3.2 *Deixis, Indexical Expressions, and Context*

Deixis refers to the way in which context enters into the interpretation of communicative exchanges, in particular, those in which reference is linguistically underdetermined. In philosophy, this complex problem is also discussed as *indexicality* (Bar-Hillel, 1954; Nunberg, 1993). A full discussion of this issue is beyond the scope of this chapter, but a brief consideration is necessary to understand a basic issue in our analysis of MUL.APIN: the relation of text to context.

Inherent in expressions of space, time, location, and motion, discussed in the previous section, is the broader issue of the relation of language to context. Spatial expressions such as *here* and *there, up* and *down*, temporal expressions like *now, then*, and *today*, motion verbs such as *come* and *go*, along with personal pronouns, such as *I, you, he*, and so on, are *deictic expressions*. They are dependent for their interpretation on a context, or frame of reference, since their meaning changes according to who is speaking, to whom, and when.

Deriving the meaning of indexical expressions from context is neither simple nor straightforward. The pronoun *you*, for example, cannot simply be defined as the addressee of an utterance, as revealed by a simple exercise in substitution (from Nunberg, 1993):

*Oh, it's you.*
*Oh, it's the addressee of this utterance.*

The first utterance conveys a range of possible interpretations, primary among these that the speaker was expecting someone else. In this case, "you" conveys both the speaker's expectation and its non-fulfillment. The second utterance does none of this work. It is, in fact, so obvious as not to be worth uttering.

Relative coordinate systems, described in the preceding section, function by invoking a *point of view*. The point of view in relation to which relative coordinate systems are understood is a particular example of a broader notion, that of a *deictic center* (Fillmore, 1997; Talmy, 2000). A deictic center is the perspective from which any deictic expression is interpreted. This perspective need not be a speaker or hearer in an oral exchange, but can also be located within a text. *Coming* or *going* can be understood relative to a character in a written (or spoken) narrative. *You* or *he* can be understood by referring back to individuals mentioned earlier in a text (or conversation). An author can also project an *authorial* perspective into the text, in which case indexical expressions are understood from the author's point of view.

The role of context, it should be emphasized, is not limited to the interpretation of indexicals. On some perspectives, context is integral to the comprehension of every communicative act, from the prelinguistic pointing of human infants (Tomasello, Carpenter & Liszkowski, 2007) to sophisticated verbal understanding on the part of intelligent adults (Sperber, 1994). On these accounts, some degree of inference is involved in every act of comprehension, inference that, by definition, appeals to non-linguistic sources of information (Sperber & Wilson, 1995).

Inferential understanding takes place within what can roughly be described as a frame of reference, variously referred to as cognitive context (Sperber & Wilson, 1995), common ground (Clark, 1996), form of life (Wittgenstein, 1955), or joint attentional frame (Tomasello et al, 2007). These terms are not strictly parallel. All involve different theoretical assumptions and are directed to understanding somewhat diverse phenomena. What they have in common is an emphasis on the importance of context in the inferential understanding of communication events.

## 3.3 *Categories and Concepts*

A primary concern of any science, not least astronomy, is the categorization, description, and definition of phenomena. The explicit, analytic process of scientific categorization is grounded in the transparent, largely automatic, universal properties of human cognition. Our minds organize our perceptions and thoughts into category structures, and necessarily so. If each object that we perceive were treated as unique, the complexity of our environment would soon overwhelm us.

Concepts are the basic constituents of human cognition (Fodor, 1994; Medin, Ross & Markman, 2001; Medin, Lynch & Solomon, 2000). Concepts enable the assignment of particulars to categories, in turn enabling a range of inferences beyond the immediately perceptible features of individuals. Concepts have both representational (semantic) and causal (relational) properties, and they enter into systems of explanation by which we understand the world and each other.

When a biologist assigns a penguin to the category *bird*, the category assignment enables a range of inferences distinct from those that would result from assigning it to the category *fish* or *mammal*, even though it shares the feature *swims under water* with cod and whales. When an astronomer, ancient or modern, assigns a heavenly body to the category *star*, he is designating a set of inferences that may be drawn.

### 3.3.1 *Kinds of Concepts*

The domain of astronomy, indeed natural science as a whole, requires a distinction between kinds of concepts. One of the basic distinctions is between *natural kinds* and *artifacts* (Medin, Lynch & Solomon, 2000). Natural kinds are objects, observed in the natural world, that take their meaning from their ontological status, such as *tiger, gold, water,* or in the case of MUL.APIN, *star.*[1] Artifacts, in contrast, are the constructed

[1] In the case of MUL.APIN, we note that domains of reference of natural kind terms are sometimes not strictly comparable to those rendered in English translation. For example, the natural kind term *star* (Sumerian MUL = Akkadian *kakkabu*), refers

products of human use or invention, and take their meaning from their function, such as *tire, chair, hat*, or in the case of MUL.APIN, a *water clock*.

The assumptions governing the distinction between these two types of phenomena, natural kinds and artifacts, are deeply-rooted and early-emerging. In the first year of life, infants demonstrate a naïve physics, expectations about the properties of objects and the forces that act upon them (Wellman & Gellman, 1992).[2] In the realm of biology, four-year-olds' inferences about living things are determined more by the essential nature of kinds than by their perceptual properties: an egg is still an egg, and a turtle is still a turtle, even when the outer shell is removed (Gelman, 2004). From an early age, then, an essentialist understanding of distinctions among natural phenomena exists. Individuating objects and assigning them to categories begins very early in life, and these basic categories form the basis for children's subsequent theorizing about science (Carey, 2000). The categorizing, defining, and theory-building of science are thus underwritten by early-emerging, basic properties of mind.

## 3.4 *Naming*

Names are integral to the formation of concepts and kinds (Kripke, 1972). To illustrate: Heraclitus claimed that "no man enters the same river twice." His observation relates to an ontological fact. The water that makes up the river is constantly changing. From the perspective of the physical universe, Heraclitus was correct, but from the perspective of language and mind, he was quite wrong. The concept named by the word *river* has a constant designation, notwithstanding the inconstancies of the physical universe. If, in some possible world, Kripke had invited Heraclitus to a debate on the bank of the river, Heraclitus would have had no problem in understanding where they were intended to meet, even if every molecule of $H_2O$ had changed since he had last set eyes on the body of water designated by the name *river*.

---

not only to individual stars as is most common in English, but also to constellations, planets, and at times to a wide range of observable astronomical phenomena.

[2] See Spelke & Kinzler, 2007; and Kinzler & Spelke, 2007, for current views on core knowledge and its development.

A name is an invitation to form a concept (Brown, 1958). Objects can be named using a variety of words or descriptions, at different levels of abstraction: *money, coin, penny, nineteen-forty-two Lincoln penny, a round piece of copper*. The name we choose may reflect our intentions toward the object, or its role in the activity, or form of life, in which we might be engaged (Wittgenstein, 1955). The names, or words, that we use to label our thoughts give us access to, and control over, our thinking. Names are to concepts what uniforms are to football players: they enable us to keep track of them in the complex play of the mind (Dennett, 1998).

Naming a scientific category can be expected to have a similar function, and thus to enhance the development of theory. Empirical studies have shown that a name for an abstract concept, such as a relational term, will enhance performance on cognitive tasks that rely on relational knowledge (Gentner, 2003). If an ambiguous green object is called a *bug*, it will lead even four-year-olds to infer that it shares more properties with other bugs than with a green leaf to which it is more perceptually similar (Gelman, 2004). Names promote the formation of object categories in human infants (Waxman & Braun, 2005). In science, names form a central role in stipulating new concepts in the process of theory formation. MUL.APIN contains many names of astronomical phenomena, and we examine their role as they appear to relate to category assignment.

## 3.5 *Definition*

Central to the practice of any science is the definition of terms and concepts. A definition renders meaning explicit. Theoretical disciplines such as mathematics and philosophy could not proceed without the aid of definition. There are up to eighteen different types of definition (Robinson, 1954), which range from a simple deictic, or ostensive, definition, which depends on the presence of the object mentioned in the immediate context:

*That is a star*

to a concise and formal definition in which a linguistic expression is intended to specify meaning precisely and completely, such as the kind of expression that forms a theorem in geometry:

*A circle is the locus of points equidistant from a given point*

The most familiar form of definition, the kind we expect to find when we look up a word in the dictionary, often takes the form of a statement of equivalence. The copula *is* links the *definiendum*, or word to be defined, with the *definiens*, an expression of its meaning:

*[word = definiendum] is [meaning = definiens]*

or, more formally, as Bierwisch and Kiefer (1969) would have it,

*NP1 is NP2*

where NP is a noun phrase. In this case, there needs to be an intrinsic semantic relation between NP1 and NP2, to distinguish definitions from statements like:

*Canada's primary export* <u>*is*</u> *timber.*

### 3.5.1 *Stipulative Definition*

A stipulative definition creates a theoretical category, as in the theorem above for circle, usually for the purpose of mathematical proof or philosophical argument, and can take the form:

*Let [word] mean [linguistic expression].*

The crucial property of stipulative definitions is that they are conceptually productive.

Much as human language is infinitely creative, so is human thought. Dennett (1994) describes the open, flexible, reflective properties of human thought as directly enabled by language, the faculty that gives human cognition an exponential advantage over that of any other species. Fodor (1994; 2001) uses a linguistic analogy, arguing that conceptual thought, like language, is componential and combinatory. Human beings can imagine, and mentally represent, an infinite number of linguistically generated concepts, even counterintuitive ones:

*a purple dog a mile long*
*a giant blue ox named "Babe"*

These linguistically generated concepts violate both our experience and our expectations, but we can think about them all the same.

The componential, generative, flexible aspect of human cognition is the basis for theory building. Science, mathematics, and philosophy accomplish their ends, to a significant degree, through the power of language to generate, or stipulate, counter-intuitive or abstract conceptual categories. The epistemological force of a stipulative definition is equivalent to that of an assumption:

> stipulative definitions are assumptions. To give a definition is to say "Let's assume for the time being that the following equivalence holds." The epistemological force of a stipulative definition is the same as the epistemological force of an assumption... "True by stipulative definition" is like "true by assumption"; just as something that is assumed to be true can turn out not to be true, something that is true by stipulative definition can turn not to be true either.
>
> Harman, 1996:399

The relation of definition to truth, and of both to science, is analyzed extensively in Davidson (1990), who differentiates between definitions that introduce new words and those aimed at expressing substantive truths, the latter more in keeping with the aims of a scientist. The intention behind a definition is crucial to understanding its force:

> Suppose we offer as a definition of the predicate "x is a solar planet" the following: x is a solar planet if and only if x is just one of the following: Mercury, Venus, Earth, Mars, Jupiter, Saturn, Uranus, Neptune, Pluto. This entails... "Neptune is a solar planet." Is this last a logical truth? One may as well say so if our definition is purely stipulative, otherwise not. The question whether it is purely stipulative is not one that can be answered by studying the formal system; it concerns the intentions of the person making the definition.
>
> Davidson, 1990:293

A definition can be purely, or merely, stipulative, on Davidson's account, in which case it is alterable according to the assumptions of a given argument; or, it can have a correspondence relation with a natural or conceptual category, in which case it is not.

Science requires exactitude, beyond that which ordinary language provides. A biologist needs to distinguish a lion from a tiger in a more precise way than a cattle farmer trying to protect his herd. An astronomer needs to distinguish among heavenly bodies in a more precise way than a casual stargazer. The need for precision in science is reflected in statements of equivalence, definitions, and generalizations over particulars. The assumptions underlying these expressions change in accordance with the advancement of knowledge.

## 3.6 *Assumptions and Axioms*

The term *axiom* has a precise and limited meaning in philosophy and mathematics. Within these disciplines, an axiom is a starting point for deduction or inference, a proposition that is held to be self-evident and that is not subject to proof or demonstration. Outside these fields, the term axiom is used more loosely, usually to refer to an established principle or generally-held assumption within an accepted body of knowledge. In our analysis of MUL.APIN, we find a number of expressions of the latter type, and at least one that borders on the former.

## 3.7 *Rhetorical Concerns*

Naturalistic analysis, as outlined above, can address individual words, simple and complex linguistic expressions, and the issue of sequence of textual components in MUL.APIN.[3] It can also bring coherence to a discussion of change or development, across textual components, by appeal to cognitive and linguistic principles that can be assumed to underlie such development. But there are some features of the text that do not appear to be governed by universal natural categories.

On our reading of MUL.APIN, several units of the treatise appear to reflect rhetorical concerns. In some cases, the opening portion of a component section refers forward to the astrological material about to be discussed. In others, a final statement seems to either summarize, or describe procedures that are to be carried out in relation to, the astronomical matter just presented. These units of text thus function in the way a modern reader might recognize as an *introduction* or a *conclusion*. There are also portions of highly-formatted text. Quantitative information, standardized measures of time and distance, are presented in a systematic and repetitive form that, to the modern reader, appears much like a table. These features of the MUL.APIN treatise seem best addressed as discourse conventions, or as rhetorical devices.

[3] See Chapter 1, 1.5, for a discussion of sequence and a cognitive perspective on sequence.

Aristotle's[4] work on the topic of rhetoric is generally recognized as seminal in the field. We are fortunate that scholars of our own era have examined modern texts, including scientific texts, in the light of classical rhetorical categories (Perelman & Olbrechts-Tyteca, 1969; Gross, Harman, & Reidy, 2002). A rhetorical analysis of the development of scientific articles from the 17th century to the present day is presented in Gross, Harman and Reidy (2002). Their analysis treats the organization of text, features such as introductions, conclusions, tables, and data displays, as features of *presentation,* while allowing that in classical rhetoric, they would probably fall under the category of *arrangement.* In our analysis, we discuss features of the MUL.APIN treatise that appear to be motivated by a concern with presentation under the heading of *rhetorical devices*.

[4] *Rhetoric* I & II

CHAPTER FOUR

# MUL.APIN: TEXT AND ANALYSIS

## *A Note on the Form of the Akkadian Text of MUL.APIN*

Although cuneiform has no formal system of punctuation marks, the scribes make use of a few devices. These include the physical unit of the line of text, the horizontal line ruling, and the simple vertical stroke (DIŠ), which marks the beginning of individual text entries. The sign DIŠ typically appears at the start of a physical line of text. Most entries in the opening sections of MUL.APIN occupy a single physical line of text, so it is normally the case that *entries* and *lines* are co-extensive. However, as the series moves on towards more complex constructions in the later sections of text, more often a single entry marked by DIŠ spans more than one physical line of text (see illustration, Plate I, lines 3–4).

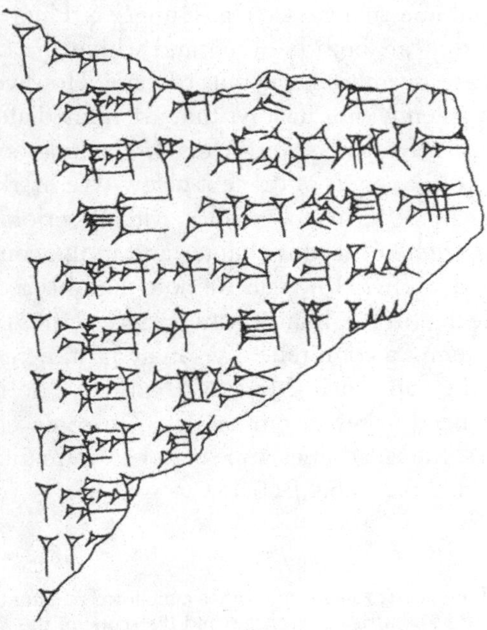

Hunger and Pingree, 1989, Plate I, F obv.

Sometimes, on the original cuneiform tablets, additional or subsequent lines are indented to indicate that they continue the entry which began above, although this is not always the case. An exception to this rule can be found in MUL.APIN at line II iii 33 (Manuscript E), the unusual situation of two DIŠ signs on a single line of text; one marking the start of a short omen occupying only the first half of the line, and the second marking the start of a short commentary on the omen. Interestingly, a second manuscript (HH) uses the cuneiform colon () to indicate that the second half of the line is indeed commentary in place of the second DIŠ, as follows:

Manuscript E: DIŠ ..................... DIŠ ..........................
Manuscript HH: DIŠ .................... : ..........................

Another example is at I ii 20 (Manuscript X) where two short lines of text, marked by DIŠ, are written on the same line of the tablet.

In our analysis, we assume that DIŠ marks units of sense that are conceptually distinct to the composers of MUL.APIN. In subsection e-1, for example, the list of fourteen *ziqpu* stars appears as a single entry marked by DIŠ, which extends over three physical lines of text, rather than as a series of separate entries (lines I iv 7–9). Entries and lines appear more discrepant from this point on in the treatise.

The horizontal line rulings (see e.g. Hunger & Pingree, 1989, plates II and III) that appear, both between and within the component sections, in some cases seem to mark units of sense. However, they are not always used consistently, nor in ways that are immediately comprehensible to the modern reader. Nevertheless, we include both the line rulings and DIŠ in our analysis of the text below. We mark DIŠ using the contemporary symbol ¶, in accordance with Assyriological tradition, both in the text segments in this chapter and in the complete text presented in Appendix One. However, we note that this is our addition, as DIŠ is not indicated in the Hunger and Pingree (1989) translation.

We do not attempt a comprehensive analysis of these "punctuation" phenomena as they appear in MUL.APIN, but we do make reference to them as the need arises in our analysis. In some cases, which we note and discuss, the ruled lines appear to bear semantic or rhetorical force (see 4.4.2.2, 4.6.2, 4.6.4.2, below).[1]

[1] The issue of line entries, rulings, and other cuneiform punctuation phenomena is worthy of a study in its own right, but is beyond the scope of this book.

## 4.1 *Section a, MUL.APIN I i 1–ii 35*

### 4.1.1 *Astronomical Content*

Section a identifies the members of a set of interest, or, the initial subject matter of the astronomical treatise that follows. It catalogues seventy-one stars,[2] divided among the three paths of the sky, identified in traditional Mesopotamian astronomy as the paths of the god Enlil (the Northern path), the god Anu (the Central path) and the god Ea (the Southern path) (Horowitz, 1998:252–258).

### 4.1.2 *Textual Form*

Section a consists of three lists that give the names, and in some cases, the location, of the stars for each of three star paths. The three star lists are similar in form. No introduction or instructions for use precede the star lists.

The first line gives the first entry. It is typical of the star lists, in that it names the star and offers an identification with the god Enlil. Typically, stars are identified with deities, and are often located in relation to other stars. All three lists are constructed according to the same pattern.

[2] Below the term 'star' is used in the Ancient Mesopotamian sense, that is to say translating Sumerian-Akkadian mul = *kakkabu*; a category which includes individual stars, constellations, and often planets and other phenomena in the sky. Thus, most of the 71 stars of MUL.APIN are in fact constellations, not individual fixed stars, and include the planets Mercury, Venus, Mars, Jupiter, and Saturn. These five, together with the Sun and Moon, comprise the category of Sumerian-Akkadian udu.idim = *bibbu* (literally, "wild sheep"), which we translate as planet following standard Assyriological convention. Thus, in the context of the Ancient Near East, one can speak of the seven planets. Uranus, Neptune, and Pluto were of course not discovered until the modern period.

### 4.1.3 *Translated Text*

Ii1 ¶ The Plow, Enlil, who goes at the front of the stars of Enlil.
Ii2 ¶ The Wolf, the seeder of the Plow.
Ii3 ¶ The Old Man, Enmešarra.
Ii4 The Crook, Gamlum.
Ii5 ¶ The Great Twins, Lugalgirra and Meslamtaea.
Ii6 ¶ The Little Twins, Alammuš and Nin-EZENxGUD (Gublaga).
Ii7 ¶ The Crab, the seat of Anu.
Ii8 ¶ The Lion, Latarak.
Ii9 ¶ The star which stands in the breast of the Lion: the King.
Ii10 ¶ The dusky stars which stand in the tail of the Lion:
Ii11 The Frond (of the date palm) of Eru, Zarpanitu.
Ii12 ¶ ŠU.PA, Enlil who decrees the fate of the land.
Ii13 ¶ The star which stands in front of it: the Abundant One, the messenger of Ninlil.
Ii14 ¶ The star which stands behind it: the Star of Dignity, the messenger of Tišpak.
Ii15 ¶ The Wagon, Ninlil.
Ii16 ¶ The star which stands in the cart-pole of the Wagon:
Ii17 The Fox, Erra, the strong one among the gods.
Ii18 ¶ The star which stands in front of the Wagon: the Ewe, Aya.
Ii19 ¶ The Hitched Yoke, the great Anu of Heaven.
Ii20 ¶ The Wagon of Heaven, Damkianna.
Ii21 ¶ The star which stands in its rope: the Heir of the Sublime Temple,
Ii22 the first-ranking son of Anu.
Ii23 ¶ The Standing Gods of Ekur, the Sitting Gods of Ekur.
Ii24 ¶ The She-Goat, Gula.
Ii25 ¶ The star which stands in front of the She-Goat: the Dog.
Ii26 ¶ The bright star of the She-Goat: Lamma, the messenger of Baba.
Ii27 ¶ The two stars which stand behind it: Nin-SAR and Erragal.
Ii28 ¶ The Panther: Nergal.
Ii29 ¶ The star which stands at its right side: the Pig, Damu.
Ii30 ¶ The star which stands at its left side: the Horse.
Ii31 ¶ The star which stands behind it: the Stag, the messenger of the Stars.
Ii32 ¶ The dusky stars which stand in the breast of the Stag:
Ii33 Harriru, the Rainbow.
Ii34 ¶ The bright red star which stands in the kidney of the Stag:
Ii35 The Deleter.
Ii36 When the stars of Enlil have been finished,
Ii37 one big star—(although) its light is dim—divides the sky in half and stands there: (that is) the star of Marduk, the Ford,
Ii38 ¶ Jupiter, (it) keeps changing its position and crosses the sky.

- - - - - - - - - - - - - - - -

Ii39 33 stars of Enlil

- - - - - - - - - - - - - - - -

Ii40 ¶ The Field, the seat of Ea, which goes at the front of the stars of Anu.
Ii41 ¶ The star which stands opposite the Field: the Swallow.
Ii42 ¶ The star which stands behind the Field, Anunitu.
Ii43 ¶ The star which stands behind it: the hired Man, Dumuzi.
Ii44 ¶ The Stars, the seven gods, the great gods.
Iii1 ¶ The Bull of Heaven, the Jaw of the Bull, the crown of Anu.
Iii2 ¶ The True Shepherd of Anu, Papsukal, the messenger of Anu and Ištar.
Iii3 ¶ The twin stars which stand opposite the True Shepherd of Anu:
Iii4 Lulal and Latarak.
Iii5 ¶ The star which stands behind it: the Rooster.
Iii6 ¶ The Arrow, the arrow of the great warrior Ninurta.
Iii7 ¶ The Bow, the Elamite Ištar, the daughter of Enlil.
Iii8 ¶ The Snake, Ningizzida, lord of the Netherworld.
Iii9 ¶ The Raven, the star of Adad.
Iii10 ¶ The Furrow, Šala, the ear of corn.
Iii11 ¶ The Scales, the horn of the Scorpion.
Iii12 ¶ The star of Zababa, the Eagle, and the Dead Man.
Iii13 ¶ Venus keeps changing its position and crosses the sky.
Iii14 ¶ Mars keeps changing its position and crosses the sky.
Iii15 ¶ Saturn keeps changing its position and crosses the sky.
Iii16 ¶ Mercury, whose name is Ninurta, rises or sets in the east
Iii17 or in the west within a month.

---

Iii18 23 stars of Anu.

---

Iii19 ¶ The Fish, Ea, who goes at the front of the stars of Ea.
Iii20 ¶ The Great One, Ea; the star of Eridu, Ea.
Iii21 ¶ The star which stands at its right: Ninmah.
Iii22 ¶ EN.TE.NA.BAR.HUM, Ningirsu.
Iii23 ¶ The star which stands at its side: The Harrow, the weapon of Mar-biti,
Iii24 inside of which one sees the subterranean waters.
Iii25 ¶ The two stars which stand behind it: Šullat and Haniš, Šamaš and Adad.
Iii26 ¶ The star which stands behind them rises like Ea
Iii27 and sets like Ea: Numušda, Adad.
Iii28 ¶ The star which stands at the left side of the Scorpion: the Mad Dog, Kusu.
Iii29 ¶ The Scorpion, Išhara, goddess of all inhabited regions.
Iii30 ¶ The Breast of the Scorpion: Lisi, Nabû.
Iii31 ¶ The two stars which stand in the sting of the Scorpion:
Iii32 Šarur and Šargaz.
Iii33 ¶ The star which stands behind them: Pabilsag.
Iii34 ¶ The Bark and the Goat-Fish.

---

Iii35 15 stars of Ea.

---

### 4.1.4 *Analysis*

The dominant feature of section a is the list format (4.1.4.1, below). Individual entries are marked as distinct units of sense in the original cuneiform text by means of the single stroke DIŠ at the start of the entry.[3] The individual entries include star names and, in some cases, their location (4.1.4.2 below). Each of the three lists ends with a taxonomic expression that names the astronomical category to which the entries belong.

| | |
|---|---|
| I i 39 | *33 stars of Enlil* |
| I ii 18 | *23 stars of Anu.* |
| I ii 35 | *15 stars of Ea.* |

#### 4.1.4.1 *Discourse Forms: List Structure*

In section a, the lists of stars identify the members of the overall set of interest that will form the primary subject matter of MUL.APIN. The three lists constitute three groups of stars associated with the three stellar paths of Enlil, Anu and Ea. One star per entry of text is usual, although, in some cases, two stars are included. In section a, single entries tend to co-occur with individual lines on the tablet, but this not always the case.

#### 4.1.4.2 *Discourse Forms: Time and Space*

In section a, some entries for individual stars include intrinsic or relative descriptions of their location. Intrinsic spatial reference expresses the position of a star(s) in relation to a star already named in the list:

I i 9 *in the breast of the Lion*

while relative spatial reference implies a particular perspective from which the star is viewed:

I i 42 *behind the Field*

[3] See above, introduction to this chapter, and Chapter 6, 6.3.3, for discussion of DIŠ.

### 4.1.4.3 *Minor Textual Form: The Planets*

Section a contains a departure from the list structure. The final entries in each of the Enlil and Anu star lists refer, not to stars, but to planets. Jupiter (Marduk) is named at the conclusion of the Enlil list:

| | |
|---|---|
| I i 36 | *When the stars of Enlil have been finished,* |
| I i 37 | *one big star—(although) its light is dim—divides the sky in half and stands there: (that is) the star of Marduk, the Ford,* |
| I i 38 | *Jupiter, (it) keeps changing its position and crosses the sky.* |

and Venus, Mars, Saturn, and Mercury appear at the end of the Anu list:

| | |
|---|---|
| I ii 13 | *Venus keeps changing its position and crosses the sky.* |
| I ii 14 | *Mars keeps changing its position and crosses the sky.* |
| I ii 15 | *Saturn keeps changing its position and crosses the sky.* |
| I ii 16 | *Mercury, whose name is Ninurta, rises or sets in the east* |
| I ii 17 | *or in the west within a month* |

These entries are distinct in that they are written in the form of complete sentences, ending, as is typical in Akkadian, with verbs. The entries for Jupiter, Venus, Mars, and Saturn contain a verb form indicating multiple and continuous acts:[4]

| | |
|---|---|
| I i 38 | *keeps changing its position* |

The entry for Mercury at the end of the Anu list is translated by Hunger and Pingree as:

| | |
|---|---|
| I ii 16–17 | *rises and sets in the east or in the west* |

However, the cuneiform terms ($^{d}$UTU.È.A = *ṣīt šamši*, $^{d}$UTU.ŠÚ.A = *ereb šamši*) literally mean "sunrise" and "sunset." They can be used to express either time at sunrise and sunset, or location, east or west, i.e., the location of sunrise and sunset. This indicates a conceptual continuity between time and space, as Mercury appears either as a morning star in the East at sunrise or as an evening star in the West at sunset.

In section a, time is referred to only in the entries for Jupiter and Mercury. Jupiter takes a relative temporal description, referring to the stars:

| | |
|---|---|
| I i 37 | *when the stars of Enlil have finished* |

---

[4] What Hunger and Pingree (1989) render in the *Gtn* stem present future.

while Mercury takes an absolute temporal description (*month*):

I ii 16–17 *Mercury, whose name is Ninurta, rises or sets in the east / or in the west within a month.*

The entries for the planets are thus more complex than those for the stars, which seems to reflect the difference in complexity between the two astronomical subclasses of phenomena being described: stars maintain a fixed position in relation to all other stars, while planets change their positions.[5]

### 4.1.5 *Categories*

In section a, each of the three star lists concludes with a summary expression, each marked off from the remainder of the text with dividing lines, above and below. In contrast to the preceding star entries, none of the summary lines are marked by the stroke DIŠ at the start of the line. Thus these summary lines name the list that precedes them, but are apparently considered by the compilers to be distinct from the list, in the same way that a title page is distinct from the contents of a book:

| | |
|---|---|
| I i 39 | *33 stars of Enlil* |
| I ii 18 | *23 stars of Anu* |
| I ii 35 | *15 stars of Ea* |

The implied structure of each of the summary statements above, while it is not explicitly stated, is thus something like:

*[all of the foregoing] are [X]*

Summary statements with similar semantic force occur later in MUL. APIN, Tablet 1 (4.4.1.5, below). In these later examples, the categorical nature of the expression is more explicit.

In section a, there is no overall summary statement at the end of the three sub-lists that would function to identify all three as one category of stars. That is, there is no statement of the form:

*71 stars of Enlil, Anu, and Ea.*

[5] See Hunger and Pingree, 1999:73 for comment on this distinction.

The compilers of MUL.APIN thus may not have recognized section a as a single unit, but rather as three independent, parallel lists which might be better understood as sections a1, a2, and a3. We have maintained Hunger and Pingree's division because the structure of the lists is so similar, and parallel sets of Anu, Enlil, and Ea star lists occur in numerous other Mesopotamian astronomical texts, in particular *The Astrolabes* (Horowitz, 2007).

## 4.2 *Sections b–d, MUL.APIN I ii 36–I iii 48*

### 4.2.1 *Astronomical Content*

Sections b, c, and d give the dates of the heliacal risings of diverse stars and the temporal relations between their risings and settings. While section a, by means of the star lists, identifies the members of the set of interest for the astronomical treatise that follows, sections b–d begin to establish the temporal parameters on their appearance.

Sections b, c, and d are identified as separate by the ancient compilers of MUL.APIN, by means of horizontal rulings drawn as dividing lines between each section. We have chosen to analyze these three sections together because of their similarity of content and form. All three describe the temporal parameters governing the appearance of the stars.

In practice, the astronomers using sections c and d would be dependent on the dates given in section b to fix the date of their observations. This interdependence also suggests that the three sections function in some sense as an ordered, unified whole.

### 4.2.2 *Textual Form*

Sections b–d consist of series of temporal expressions. Each section is distinguished by a particular repeating syntactic form in which temporal relations among the individual fixed stars are expressed.

### 4.2.3 *Translated Text*

Section b gives fixed, calendar dates for the heliacal rising of various stars:

Iii36 ¶ On the 1st of Nisannu the Hired man becomes visible.
Iii37 ¶ On the 20th of Nisannu the Crook becomes visible.
Iii38 ¶ On the 1st of Ajjaru the Stars become visible.
Iii39 ¶ On the 20th of Ajjaru the Jaw of the Bull becomes visible.
Iii40 ¶ On the 10th of Simanu the True Shepherd of Anu and the Great Twins become visible.
Iii41 ¶ On the 5th of Du'uzu the Little Twins and the Crab become visible.
Iii42 ¶ On the 15th of Du'uzu the Arrow, the Snake, and the Lion
Iii43 become visible; 4 minas is a daytime watch, 2 minas is a nighttime watch.
Iii44 ¶ On the 5th of Abu the Bow and the King become visible.
Iii44a ¶ On the 1st of Ululu [....][6]
Iii45 ¶ On the 10th of Ululu the star of Eridu and the Raven become visible.
Iii46 ¶ On the 15th of Ululu ŠU.PA, Enlil, becomes visible.
Iii47 ¶ On the 25th of Ululu the Furrow becomes visible.
Iiii1 ¶ On the 15th of Tešritu the Scales, the Mad Dog, EN.TE.NA.BAR.HUM
Iiii2 and the Dog become visible; 3 minas is a daytime watch, 3 minas is a nighttime watch.
Iiii3 ¶ On the 5th of Arahsamnu the Scorpion becomes visible.
Iiii4 ¶ On the 15th of Arahsamnu the She-Goat and the Breast of the Scorpion become visible.
Iiii5 ¶ On the 15th of Kislimu the Panther, the Eagle
Iiii6 and Pabilsag become visible.
Iiii7 ¶ On the 15th of Ţebetu SIM. MAH, (i.e.) the Swallow (or) IM. ŠEŠ,
Iiii8 becomes visible in the East, and the Arrow
Iiii9 becomes visible in the evening; 2 minas is a daytime watch, 4 minas is a nighttime watch.
Iiii10 ¶ On the 5th of Šabaţu the Great One, the Field, and the Stag become visible.
Iiii11 ¶ On the 25th of Šabaţu Anunitu becomes visible.
Iiii12 ¶ On the 15th of Addaru the Fish and the Old Man become visible.

Section c gives a conjunction of the events, or the co-temporal rising and setting of the diverse stars:

Iiii13 ¶ The Stars rise and the Scorpion sets.
Iiii14 ¶ The Scorpion rises and the Stars set.
Iiii15 ¶ The Bull of Heaven rises and ŠU.PA sets.

[6] This appears as an extra line in only two manuscripts and is not numbered in the Hunger-Pingree edition.

Iiii16 ¶ The True Shepherd of Anu rises and Pabilsag sets.
Iiii17 ¶ The Arrow, the Snake, and the Lion rise, and
Iiii18 the Great One and the Eagle set.
Iiii19 ¶ The Bow and the King rise, and the She-Goat sets.
Iiii20 ¶ The star of Eridu and the Raven rise, and the Panther sets.
Iiii21 ¶ ŠU.PA, Enlil, rises and the Field sets.
Iiii22 ¶ Ninmah rises and Anunitu sets.
Iiii23 ¶ The Scales, the Mad Dog, and EN.TE.NA.BAR.HUM
Iiii24 rise, and the Hired Man sets.
Iiii25 ¶ The Scorpion and the Dog rise, and the Star of Eridu and the Stars set.
Iiii26 ¶ The Breast of the Scorpion and the She-Goat rise, and the Old Man and the True Shepherd of Anu set.
Iiii27 ¶ Pabilsag, Zababa, and the Standing Gods rise, and
Iiii28 the Arrow, the Bow and the Crook set.
Iiii29 ¶ The Panther and the Eagle rise, and
Iiii30 the Great Twins and the Little Twins set.
Iiii31 ¶ The Field, the Great One and the Stag rise and
Iiii32 the Lion, the Snake, and EN.TE.NA.BAR.HUM set.
Iiii33 ¶ The Fish and the Old Man rise, and the Furrow and the Mad Dog set.

---

Section d gives the number of days that pass between the risings of individual stars:

Iiii34 ¶ 55 days pass from the rising of the Arrow to the rising of the star of Eridu.
Iiii35 ¶ 60 days pass from the rising of the Arrow to the rising of ŠU.PA.
Iiii36 ¶ 10 days pass from the rising of ŠU.PA to the rising of the Furrow.
Iiii37 ¶ 20 days pass from the rising of the Furrow to the rising of the Scales.
Iiii38 ¶ 30 days from the rising of the Scales to the rising of the She-goat.
Iiii39 ¶ 30 days pass from the rising of the She-Goat to the rising of the Panther.
Iiii40 ¶ 30 days pass from the rising of the Panther to the rising of the Swallow.
Iiii41 ¶ 20 days pass from the rising of the Swallow to the rising of the Field.
Iiii42 ¶ 40 days pass from the rising of the Field to the rising of the Fish.
Iiii43 ¶ 35 days pass from the rising of the Fish to the rising of the Crook.
Iiii44 ¶ 10 days pass from the rising of the Crook to the rising of the Stars.
Iiii45 ¶ 20 days pass from the rising of the Stars to the rising of the Bull of Heaven.
Iiii46 ¶ 20 days pass from the rising of the Bull of Heaven to the rising of the True Shepherd of Anu.

Iiii47 ¶ 35 days pass from the rising of the True Shepherd of Anu to the rising of the Arrow.
Iiii48 ¶ 20 days pass from the rising of the Arrow to the rising of the Bow.

---

### 4.2.4 *Analysis*

#### 4.2.4.1 *Discourse Forms: Time and Space*

Sections b, c, and d together set the temporal parameters on the appearance of the stars named in section a.

##### 4.2.4.1.1 *Discourse Forms: Section b (MUL.APIN I ii 36–I iii 12)*

In section b, temporal parameters are established in relation to an absolute, or fixed, external criterion, the ideal calendar of 360 days.[7] Section b is a series of complete sentences, each beginning with a calendar date, followed by the name of the star which becomes visible on that date:

*On (X date) star (Y) becomes visible.*

##### 4.2.4.1.2 *Discourse Forms: Section c (MUL.APIN I iii 13–I iii 33)*

Section c is a series of parallel relative expressions. The rising of one star is described in relation to the setting of another star:

*X rises and Y sets.*

The highly temporal nature of the discourse in section c, in which the individual entries express two simultaneous, or nearly simultaneous, occurrences, is marked in Akkadian by the suffix *-ma* (*star X rises-ma star Y sets*) that is attached in each case to the verb 'rises'. In such constructs the *-ma* has the force of English *and* or *and then*, in the sense of two immediately consecutive actions. When a simple conjuction between two nouns is required, as in the case when more than one star rises or sets, the Akkadian conjunction *u* (and) is used:

*Star 1 u Star 2 rise -ma Star 3 u Star 4 set.*

---

[7] For the 360-day year, see Hunger & Pingree, 1989:39–40, and more recently, Brack-Bernsen, 2007.

4.2.4.1.3 *Discourse Forms: Section d (MUL.APIN I iii 34–I iii 48)*
Section d is a series of expressions that refer to both absolute and relative coordinating systems. Each entry specifies the number of days that pass (absolute measure), but the number is determined relative to the rising of the previous star. Only risings, and not settings, are referred to, but the expressions are arguably more complex than section **c**, using a "from / to" structure:

*(# of days) pass from X to Y*

4.2.4.1.4 *Minor Textual Form in Section b*
When the calendrical dates associated with solstices and equinoxes are mentioned, the compilers of MUL.APIN add a statement regarding the length of day and night, as measured in units of *minas* on a water clock (Brown, Fermor & Walker, 1999), as the following two lines of text illustrate:

| | |
|---|---|
| I ii 42 | *On the 15th of Du'uzu, the Arrow, the Snake, and the Lion become visible;* |
| I ii 43 | *4 minas is a daytime watch, two minas is a nighttime watch.* |

In this expression, we see the first appearance of the copula "is" in the translation. The copula is only implicit in the original Akkadian, a pattern typical of Semitic languages in the present tense; for example, the expression "A B," with no explicit marking of the form "to be," conveys the sense "A is B." In Akkadian, as in Hebrew and Arabic, "He boy" would convey the sense, "He is a boy."

These lines of section b thus have the same semantic force as "statements of equivalence" between two stipulated, or non-natural categories (see 3.5.1; 4.2.5): a "daytime watch" and "4 minas."

We see this general form repeated twice more in section b, once for the fall equinox (I iii 2) and once for the winter solstice (I iii 7–9):

| | |
|---|---|
| I iii 2 | *and the Dog become visible; 3 minas is a daytime watch, 3 minas is a nighttime watch.* |
| I iii 7 | *On the 15th of Ṭebetu SIM. MAH, (i.e.) the Swallow (or) IM. ŠEŠ,* |
| I iii 8 | *becomes visible in the East, and the Arrow* |
| I iii 9 | *becomes visible in the evening; 2 minas is a daytime watch, 4 minas is a nighttime watch.* |

In the entry for the winter solstice, we also see additional references to time and space: in the East (I iii 8), and in the evening (I iii 9).

The compilers do not, however, give this form for the spring Equinox, which would have been expected for the 15th of Nisannu, as no star is associated with this date in this section of MUL.APIN.

### 4.2.5 *Categories*

Sections b–d deal primarily with types of events as opposed to the simpler object categories that formed the primary content of section a; here, there is no summary category statement. These sections also show a transition from natural, or observed categories to constructed, or artifact categories (3.3). In other words, section a mentions stars and planets, while sections b–d mentions *mina*, a measure of time determined by a constructed artifact, a water clock. Here, *mina* is used to define the length of a "daytime watch," in a statement of equivalence (3.5).

## 4.3 *Intermediate Section, MUL.APIN I iii 49–50*

Between Hunger and Pingree's sections d and e is a two-line section that is not considered in their summary of the structure of MUL. APIN, presumably because the lines seem anomalous.[8]

### 4.3.1 *Astronomical Content*

The intermediate section describes the parameters of the apparent rotation of the celestial sphere in relation to earth. The textual form of the intermediate section is comprised of two complete sentences in the form of generalized descriptions.

### 4.3.2 *Translated Text*

| | |
|---|---|
| Iiii49 | The stars enter into the night in the morning 1 UŠ each day. |
| Iiii50 | The stars come out into the day in the evening 1 UŠ each day. |

---

[8] This two-line statement is also found in a *ziqpu* star text. See Horowitz 1994:93.

### 4.3.3 *Analysis*

#### 4.3.3.1 *Discourse Forms: Time and Space, Generalized Description*

The intermediate section expresses generalizations over the set of stars as a whole.

Two statements describe the parameters governing the appearance of the stars using a fixed, absolute measure of distance through space, the degree, 1 UŠ. Section a identified the members of the set, sections b–d describe parameters governing the individual members of the set; the intermediate section gives parameters over the set as a whole.

The two statements take the form of generalizations, in that the propositions they express are held to be generally true, or to hold across occasions of use. It is significant that these two entries are not marked by DIŠ, suggesting that the authors of MUL.APIN recognized that they are of a different nature than the previous entries. The only other entries up until this point not marked by DIŠ are the summary statements at the end of the lists of the stars of the paths of Anu, Enlil, and Ea. From this point on, we find numerous examples of similar distinctions between lines marked by DIŠ and those unmarked by DIŠ.

It could be argued that the entries referring to individual stars encountered in sections a through d are also general, in that they hold across different occasions of observation. That is, the names of the stars and the temporal parameters that govern their appearance are not one-time occurrences, but rather predictable, annual recurring events. However, in the intermediate section discussed here (I iii 49–50) the generalization is over the set, or category of stars as a whole, rather than over individuals.

#### 4.3.3.2 *Rhetorical Device: Proto-Axioms*

The two expressions in the intermediate section can be construed as statements of accepted, or axiomatic, assumptions (see 3.6). Minimally, they can be construed as proto-axioms. Much of traditional Mesopotamian astronomical theory is based on the assumption of a 360-day astronomical year, and the circular nature of astronomical movements. Insofar as these statements presuppose knowledge of the circular model of stellar motion, whereby the annual pattern of the movement of the stars shifts one degree per night over an ideal stellar year of 360 days, they can be construed as axiomatic.

#### 4.3.3.3 *Rhetorical Function: Transition*

We suggest that this two-line section is best understood as a transition between the simple stellar material that appears in sections a through d and the more complex discourse that appears in the following sections, beginning with section e. The two statements in this intermediate section also occur in the *ziqpu* star text (Horowitz, 1994:93). This is significant, in that here in MUL.APIN they function, in some sense, as an introduction to the description of *ziqpu* stars in section e (below).

### 4.3.4 *Categories*

The intermediate section introduces another stipulated category: a fixed, absolute measure of distance through space, the degree, 1 UŠ. As with *mina*, above (see 4.2.4.1.4), UŠ marks a conceptual transition from observed, natural kind categories (stars and planets) to a constructed, or in this case stipulated (see 3.5.1), measure of distance, a degree.

## 4.4 *Section e, MUL.APIN I iv 1–30*

Section e details the *ziqpu* stars.[9] Hunger and Pingree, as well as the ancient compilers of MUL.APIN, treat these lines as a single section. Our analysis of section e suggests a division into two subsections which we refer to below as subsections e-1 and e-2.

### 4.4.1 *Subsection e-1, MUL.APIN I iv 1–9*

#### 4.4.1.1 *Astronomical Content*

Subsection e-1 describes the *ziqpu* stars in the path of Enlil.

#### 4.4.1.2 *Textual Form*

The language of subsection e-1 begins to approach the form of continuous flowing discourse. The logic of the expressions crosses the

---

[9] For this category of stars, see Hunger and Pingree, 1999, 84–90.

line boundaries of the cuneiform text. The length of lines seems to be determined more by distinct units of sense, or conceptual units, rather that by the physical limitations of stylus on clay. In other words, entries and physical lines do not strictly co-occur, as was the norm in sections a through d of MUL.APIN.

Subsection e-1 begins with a three-line introduction that gives the procedure for observing *ziqpu* stars, followed by three lines which name fourteen *ziqpu* stars. The subsection ends with a three-line summary which is nearly identical to the introduction, and so serves as an *inclusio.* The textual form of subsection e-1, taken as a whole, is thus far more developed than the earlier sections of MUL.APIN. Rather than employing a list structure, it is characterized by continuous discourse as well as by new rhetorical devices.

### 4.4.1.3 *Translated Text*

Iiv1 ¶ The *ziqpu* stars which stand in the path of Enlil in the middle of the sky
Iiv2 opposite the breast of the observer of the sky, and
Iiv3 by means of which he observes the rising and setting of the stars at night (are the following):

---

Iiv4 ¶ ŠU.PA, the star of Dignity, the Standing Gods, the Dog,
Iiv5 the She-Goat, the Panther, the Stag, the Old Man, the Crook,
Iiv6 the Great Twins, the Crab, the Lion, Eru, and the Abundant One.

---

Iiv7 All these are the *ziqpu* stars in the path of the stars of Enlil which stand in the middle of the sky
Iiv8 opposite your breast, and by means of which you observe
Iiv9 the risings and settings of the stars at night.

---

### 4.4.1.4 *Analysis*

#### 4.4.1.4.1 *Rhetorical Devices: Introduction and Conclusion*

Subsection e-1 begins with an introduction. Hunger and Pingree add in line I iv 3 "(*are the following*)" at the end of the introduction to allow the modern reader to make sense of the statement, thus suggesting that a complete sentence is implied. We should note, however, that the original Akkadian text does not offer a syntactically complete sentence.

Subsection e-1 ends with a summary statement, in the form of a complete sentence. Rhetorically, this statement functions as a conclusion.

The use of DIŠ in subsection e-1 is instructive as to how the compilers of MUL.APIN may have understood the function of the introduction and conclusion. DIŠ is used once to open the subsection in I iv 1, and once in I iv 4 to open the three-line entry that contains the star names. But no DIŠ appears at the start of the conclusion. Based on the patterns observed in section a, this may indicate that the conclusion in subsection e-1 functions as a summary, as it is unmarked in a manner similar to the star list summaries (4.1.5). Another intriguing possibility is that the DIŠ at the start of e-1 in a sense governs the entire section, with the DIŠ at the start of the stars entry as a sort of "sub-DIŠ." Perhaps for the same reason, there is no DIŠ at the start of the next piece of the *ziqpu* star discussion, in I iv 10 (4.4.2, below).

*4.4.1.4.2 Rhetorical Devices: Direct Address*

The introduction to subsection e-1 also includes the rhetorical device of direct address. Apparently assumed to be an astronomer, the user of the text is initially referred to as:

I iv 2 *the observer of the sky*

The "*observer*" is also the implied subject of the third person masculine singular verb in the relative clause which forms the second half of the introduction:

I iv 3 *by means of which he observes*

The "*observer*" is also implied in the conclusion, where direct address is indicated by the pronominal second person, "*you*" and "*your*." The fact that the section opens in third person and then transitions to a second-person direct address may be of some significance (see 5.3).

4.4.1.4.3 *Discourse Devices: Continuous Discourse*

In subsection e-1, the *ziqpu* star names are presented in one continuous, sequential expression that spans the physical lines of cuneiform text. Between the introduction and summary sentences, the star names appear like a single complex subject of a sentence, or a single unit of sense, marked by DIŠ at the beginning of the entry, before the first star name only (see 4.4.1.5, below):

*star name 1, star name 2, star name 3*

However, there is no verb. As this complex subject does not form part of a complete sentence, it is arguably still "list-like," but unlike earlier star lists in section a, there is no elaboration of the individual entries for the stars.

4.4.1.4.4 *Discourse Forms: Space and Time, Multiple Marking*

In subsection e-1, for the first time in MUL.APIN, the stars are explicitly located relative to the position of the "*observer*," or user of the text. This serves to identify the deictic center of the discourse (see 3.2); that is, it identifies the perspective from which the stars are expected to be viewed: from the position of the astronomer, *the observer of the sky*.

Expressions describing the location of the *ziqpu* stars are more complex than those appearing earlier in MUL.APIN. Three locative expressions, two intrinsic and one relative, are found in one single long and complex entry, consisting of the lines I iv 1–3:

| | |
|---|---|
| I iv 1 | DIŠ...in the path of Enlil in the middle of the sky |
| I iv 2 | opposite the breast of the observer of the sky... |
| I iv 3 | .... |

4.4.1.4.5 *Generalizations*

The introductory and concluding sections of subsection e-1 are expressed in general form.

4.4.1.5 *Categories*

In subsection e-1, as noted in 4.4.1.4.3 above, the star names appear as a single unit of sense, marked by the stroke DIŠ at the beginning of the entry, which continues over multiple lines. Thus, the group of stars appears as a single unit, this being the category of *ziqpu* stars.

The summary statement reinforces this notion. It explicitly names the category to which the foregoing stars belong:

| | |
|---|---|
| I iv 7 | *All these are the ziqpu stars* |

4.4.2 *Subsection e-2, MUL.APIN I iv 10–30*

4.4.2.1 *Astronomical Content*

Subsection e-2 is concerned with two simultaneous observations: first, the *ziqpu* of a *ziqpu* star, in the middle of the sky (that is, the zenith of the star in its arc across the sky); second, the heliacal rising of a star, or stars, on the eastern horizon.

#### 4.4.2.2 *Textual Form*

Subsection e-2 begins with a set of instructions, or procedures, regarding where and how to stand in order to observe the *ziqpu*. These instructions form the necessary basis for making all subsequent observations over the course of the year. This introduction is followed by a list of stellar observations associated with different days and months of the year. The opening instructions form a kind of introduction, but unlike subsection e-1, subsection e-2 does not end with a summary, or *inclusio*.

It is noteworthy that the compilers of MUL.APIN utilize horizontal dividing lines to separate subsection e-2 into component entries, all marked by DIŠ, with the exception of the first five-line component.

#### 4.4.2.3 *Translated Text*

| | |
|---|---|
| Iiv10 | If you are to observe the *ziqpu*, you stand |
| Iiv11 | in the morning before sunrise, West to your right, |
| Iiv12 | East to your left, your face directed towards South; |
| Iiv13 | on the 20th of Nisannu the *kumāru* of the Panther stands in the middle of the sky |
| Iiv14 | opposite your breast, and the Crook rises. |
| Iiv15 | ¶ On the 1st of Ajjaru, the Breast of the Panther stands in the middle of the sky |
| Iiv16 | opposite your breast, and the stars rise. |
| Iiv17 | ¶ On the 20th of Ajjaru, the Knee of the Panther |
| Iiv18 | stands in the middle of the sky opposite your breast, and the Jaw of the Bull rises. |
| Iiv19 | ¶ On the 10th of Simanu the Heel of the Panther stands in the middle of the sky |
| Iiv20 | opposite your breast, and the True Shepherd of Anu rises. |
| Iiv21 | ¶ On the 15th of Du'uzu the bright star of the Old Man stands in the middle of the sky opposite your breast, and the Arrow rises. |
| Iiv22 | ¶ On the 15th of Abu the dusky stars of the Old Man |
| Iiv23 | stand in the middle of the sky opposite your breast, and the Bow rises. |
| Iiv24 | ¶ On the 15th of Ululu the Great Twins stand in the middle of the sky opposite your breast, and ŠU.PA and the Star of Eridu rise. |
| Iiv25 | ¶ On the 15th of Tešritu the Lion stands in the middle of the sky opposite your breast, and the Scales rise. |

| | |
|---|---|
| Iiv26 | ¶ On the 15th of Arahsamnu Eru stands in the middle of the sky opposite your breast, and the She-Goat rises. |
| Iiv27 | ¶ On the 15th of Kislimu ŠU.PA stands in the middle of the sky opposite your breast, and the Panther rises. |
| Iiv28 | ¶ On the 15th of Ṭebetu the Standing Gods stand in the middle of the sky opposite your breast, and the Swallow rises. |
| Iiv29 | ¶ On the 15th of Šabaṭu the Dog stands in the middle of the sky opposite your breast, and the Field rises. |
| Iiv30 | ¶ On the 15th of Addaru the She-Goat stands in the middle of the sky opposite your breast, and the Fish rises. |

#### 4.4.2.4 *Analysis*

##### 4.4.2.4.1 *Rhetorical Devices: Introduction, Direct Address*

Subsection e-2 begins with an introduction (lines I iv 10–14) that gives explicit instructions to the user of the text. It employs the second-person form of address, "*you.*" The instructions begin with the conditional marker "*if*" combined with the infinitive form of the verb:

> "*If you are to observe...*"

In the succeeding series of month-by-month stellar observations, subsection e-2 repeatedly uses second-person direct address "*you*" or "*your,*" the implied reference being "*the observer of the sky*" in each entry in the section.

##### 4.4.2.4.2 *Dividing Lines*

The first five lines in e-2 (I iv 10–14), shown in 4.4.2.3 above, precede the first dividing line. Lines I iv 10–12 give instructions, or procedures; while lines I iv 13–14 give the first set of observations for the month of Nisannu. These lines appear to be taken as a single unit by the compilers of MUL.APIN with all five lines unmarked by DIŠ, perhaps because as we suggested above (4.4.1.4.1) the DIŠ in line I iv 1 governs the entire *ziqpu* star discussion.

Surprisingly, none of the different versions of the MUL.APIN tablets place a horizontal dividing line between the opening instructions and the first set of observations, those for the first month Nisannu. This might seem to be a logical place for a line, as it would serve to separate the procedure from the first piece of observational data. In the remainder of subsection e-2, dividing lines occur between each subsequent set of astronomical observations, for each separate month, each of which opens with DIŠ, indicating the start of an entry.

This suggests that the compilers of MUL.APIN considered the first five lines of e-2 to be a single unit of text. Thus, this opening textual unit of e-2 can be characterized as an introduction which includes both instructions and the first set of resulting observations. This is interesting from a scientific perspective, as it links the procedure and the first instance of its application as a conceptual unit.

#### 4.4.2.4.3 *Discourse Forms: Space and Time, Multiple Marking*

Subsection e-2 uses absolute, relative, and intrinsic coordinate systems. The entries give multiple markings of the parameters governing the appearance of the stars. Similar structures were noted in subsection e-1 (4.4.1.4.4 above), but the locative expressions in subsection e-2 are even more complex.

The entries begin with a calendar date, followed by the name of the star which rises on the date, and the cardinal directions east and west. They use distal intrinsic expressions which are repeated in the various entries:

> *in the middle of the sky*

expressions relative to the position of the observer:

> *opposite your breast*

and expressions that use both relative and absolute reference together:

> *your face directed toward the South*

Five locative references are included in a single entry in I iv 10–13:

> *West to your right*
> *East to your left*
> *Your face directed towards South*
> *In the middle of the sky*
> *Opposite your breast*

These are to some extent redundant, or "over-marked," in that if one's face is directed toward the South, West is necessarily to your right, etc. Similarly, there are also multiple temporal expressions in this same entry; both relative (before sunrise) and absolute (calendar date):

*in the morning*
*before sunrise*
*on the 20th of Nisannu*

#### 4.4.2.4.4 *Generalizations*

Subsection e-2 contains complex generalized expressions that describe particular astronomical arrays associated with specific dates, arrays that can be expected to regularly recur. The expressions take the following, repeated form:

*On (X date), (Y) is observed.*

#### 4.4.2.5 *Categories*

There is no summary statement in subsection e-2 that names a category. This, in all likelihood, is a consequence of the content. Subsection e-2 gives a series of parallel procedural observations in sequence over the course of the year. It is not a list of objects, such as stars. Unlike stars, a series of complex observations does not yield easily to taxonomic assignment. However, the repetitive formalistic structure of the entries is suggestive of an incipient or inchoate procedural type, or category.

### 4.5 *Section f, MUL.APIN I iv 31–II i 8*

The moon figures prominently in all of the material in section f. This likely motivated Hunger and Pingree to include the last lines of MUL. APIN Tablet 1 and the first lines of MUL.APIN Tablet II in a single section, which they call "The Path of the Moon."

However, other considerations suggest that two separate sections of text are indicated. We shall refer to these as subsections f-1 and f-2. The justification for the division is threefold:

*Division Justification 1: Physical Separation*
Subsection f-1 closes Tablet I and subsection f-2 opens Tablet II. The two subsections of text were physically separated by the ancient compilers of MUL.APIN in the standard two-tablet format of the series (see 1.2). Thus, the two subsections were not strictly consecutive in the standard canonical form of the ancient cuneiform treatise, which suggests that they were considered separate entities.

Also, the colophons at the ends of exemplars of Tablet I would have provided a further physical separation between the last line of Tablet I and the first line of Tablet II in the cuneiform originals.[10]

*Division Justification 2: Astronomical Content*

Subsection f-1 names the stars in the path of the moon. Subsection f-2 names the planets (Akkadian *bibbu*) that travel the same path as the moon travels; the planets being, for the Mesopotamians, the moon itself and the six other planets listed in f-2. Hence, f-1 shares the dominant concern with stars of MUL.APIN Tablet I while f-2 does not.

*Division Justification 3: Textual Form*

The textual form of subsection f-1 at the end of Tablet I, is more consistent with the earlier sections of MUL.APIN that appear on Tablet I, while the textual form of subsection f-2 on Tablet II, is more consistent with later sections found on Tablet II (see section g, below, the first intercalation scheme, which illustrates this change). In this sense, subsection f-2 appears to serve as a transitional section. Hence, there is logic to treating f-1 and f-2 separately, although they are designated as a single section f in Hunger and Pingree (1989).

### 4.5.1 *Subsection f-1, MUL.APIN I iv 31–39*

#### 4.5.1.1 *Astronomical Content*

Subsection f-1 consists of names of the stars in the path of the Moon.

[10] Colophons are placed by scribes at the bottom of the reverse of cuneiform tablets separated by a single or double horizontal dividing line from the main text. They function as a sort of title page, typically giving in sequence many of the following components: the catchline of the tablet (= the first line of the next tablet); the title and tablet number of the current tablet (for example Tablet II of MUL.APIN), the name of the scribe, his rank, and family affiliation, the date, and sometimes other miscellaneous information. Hunger, 1968, is a monograph-length study of cuneiform colophons with a detailed introduction to the subject. The six surviving colophons from sources for MUL.APIN are published in Hunger and Pingree 1989:123.

#### 4.5.1.2 *Textual Form*

Subsection f-1 is similar to the *ziqpu* star texts discussed above (subsection e-1). It consists of an introduction, star names, and a concluding statement.

#### 4.5.1.3 *Translated Text*

| | |
|---|---|
| Iiv31 | The gods who stand in the path of the Moon, through whose regions the Moon in the course of a month |
| Iiv32 | passes and whom he touches: |
| Iiv33 | The Stars, the Bull of Heaven, the True Shepherd of Anu, the Old Man, |
| Iiv34 | the Crook, the Great Twins, the Crab, the Lion, |
| Iiv35 | the Furrow, the Scales, the Scorpion, Pabilsag, |
| Iiv36 | the Goat-Fish, the Great One, the Tails, the Swallow, |
| Iiv37 | Anunitu, and the Hired Man. |
| Iiv38 | All these are the gods who stand in the path of the Moon, through whose regions the Moon in the course of a month |
| Iiv39 | passes and whom he touches. |

#### 4.5.1.4 *Analysis*

##### 4.5.1.4.1 *Rhetorical Devices: Introduction and Conclusion*

As in the *ziqpu* star texts subsection e-1, subsection f-1 also begins with an introduction and ends with a summary statement, or conclusion. Here, f-1 differs from e-1 in that there is no direct address of the user of the text, nor reference to specific procedures or goals.

##### 4.5.1.4.2 *Discourse Forms: Time and Space, Complex Descriptions*

Both the introduction and conclusion sections in subsection f-1 use multiple pronominal and prepositional phrases (I iv 31-2) that refer to the moon by means of intrinsic (in the path of the moon) and relative (whom he touches) spatial coordinate systems:

*(who stand) in the path of the Moon*
*through whose regions (the Moon...passes)*
*whom he touches*

Temporal descriptions use an absolute coordinate system to express the interval of time within which the movement of the stars is observed.

This measure of time is, not surprisingly, the month, which is defined by the moon's phases in the lunar calendar:

I iv 31 ... *in the course of a month*

#### 4.5.1.4.3 *Generalizations*

Both the introduction and conclusion in subsection f-1 are complete complex sentences, expressed in generalized form, referring to recurring observable states of affairs.

### 4.5.1.5 *Categories*

Similar to the *ziqpu* star texts in section e, the star names are presented in a continuous format rather than a list of separate entries. This textual form groups the stars as a single complex subject in the text, or a single conceptual unit. Both the introduction and summary statements in subsection f-1 are descriptions over the set of stars, but subsection f-1 differs from earlier sections in referring to the set of stars collectively, in both the introduction and the conclusion, as "*the gods*":

I iv 31 *the gods who stand in the path of the Moon....*
I iv 38 *all these are the gods who stand in the path of the Moon....*

This is in contrast to the earlier *ziqpu* star texts, which refer to individual stars as gods but to the set or category of stars collectively as "*stars.*"

However, we note that in the middle portion of section f-1 (I iv 33–37) the authors write the actual names of these stars in the Akkadian original using the star determinative (MUL), rather than using the god determinative (DINGIR). Stars in Mesopotamia were considered to be gods, and constellations to be the heavenly images of important gods, and thus the categories "stars" and "gods" are not mutually exclusive. In MUL.APIN and elsewhere, the names of the Moon god and Sun god (i.e. the Moon and the Sun) are written with the god determinative (DINGIR).

## 4.5.2 *Subsection f-2, MUL.APIN II i 1–8*

### 4.5.2.1 *Astronomical Content*

Subsection f-2 describes the planets (*bibbu*) that are to be found in the path of the Moon.

#### 4.5.2.2 *Textual Form*

Subsection f-2 is comprised entirely of complete sentences, a series of parallel syntactic structures that describe the paths of the planets, followed by a concluding summary statement. It may be worth noting here that Hunger and Pingree's translation of the first six sentences alters somewhat the syntactic sequence of the original Akkadian. A more faithful translation might begin f-2 as follows:

> *the path that the Moon travels, the Sun travels.*
> *the path that the Moon travels, Jupiter travels [etc.]*

#### 4.5.2.3 *Translated Text*

| | |
|---|---|
| IIi1 | ¶ The Sun travels the (same) path the Moon travels. |
| IIi2 | ¶ Jupiter travels the (same) path the Moon travels. |
| IIi3 | ¶ Venus travels the (same) path the Moon travels. |
| IIi4 | ¶ Mars travels the (same) path the Moon travels. |
| IIi5 | ¶ Mercury whose name is Ninurta travels the (same) path the moon travels. |
| IIi6 | ¶ Saturn travels the (same) path the Moon travels. |
| IIi7 | Together six gods who have the same position, (and) who touch the stars of the sky and |
| IIi8 | keep changing their positions. |

#### 4.5.2.4 *Analysis*

##### 4.5.2.4.1 *Rhetorical Devices: Conclusion*

Subsection f-2 concludes with a summary statement that describes the planets as a set. This is another example of the apparent convention of not marking summary statements by DIŠ.

##### 4.5.2.4.2 *Discourse Forms: Space and Time*

The character of the descriptions in subsection f-2 differs from those in earlier sections. Paths, rather than positions, of the astronomical objects are described. This difference reflects the fact that the objects being described are planets rather than fixed stars. Paths of the planets are described relative to the path of the Moon.

##### 4.5.2.4.3 *Generalizations*

Subsection f-2 is a series of six complete sentences with parallel syntactic forms, each describing the path of one of the six ancient planets (including the Sun), all of which "travel the same path" as the Moon (in Mesopotamian cosmology, the Moon was considered the seventh planet (see 4.1.1, footnote 2). The sentences are each generalized descriptions of observations that would be expected to recur.

### 4.5.2.5 *Categories*

The summary statement of subsection f-2 is similar to that in f-1 (see 4.5.1.4.1): it categorizes the planets as "*the six gods*." The unique property that distinguishes planets (*they keep changing their position*) from stars is noted explicitly (see also 4.1.4.3, above, where the property was predicated of the individual planets Jupiter and Mercury, whereas here, it is a generalization over the set of planets):

II i 8 *(they) keep changing their positions*

In contrast, the stars maintain fixed positions relative to one another. The concluding statement thus distinguishes properties of the set being described (the planets) from the set previously identified (stars).

## 4.6 *Section g, MUL.APIN II i 9–24*

### 4.6.1 *Astronomical Content*

Section g details the equinoxes and solstices. Hunger and Pingree refer to this section as the first intercalation scheme, since its purpose is the reconciling (by intercalation) of discrepancies between different reckonings of the year. The non-intercalated civil calendar of 12 lunar months, which has 354 days, loses eleven-plus days per year to the true stellar/solar year of 365 1/4 days. The information in this text would have been used by ancient astronomers to determine the number of discrepant days, with the objective of bringing the lunar year back into alignment with the stellar/solar year.

### 4.6.2 *Textual Form*

Section g is a series of complete sentences, complex general statements that use multiple markings of the temporal and spatial parameters governing the appearance of heavenly bodies. The concluding statement summarizes the dates of the solstices and equinoxes, and serves to identify the foregoing as a single unit of text.

The ancient compilers use horizontal line rulings in this section of text to set off five subsections in accordance with the astronomical content. Each of these subsections is marked by DIŠ and in this sense, each can be considered an entry. The first four entries are descriptions of the solstices and equinoxes, in the following order: Summer solstice, Fall equinox, Winter solstice, Spring equinox. The final line ruling sets apart the summary statement.

### 4.6.3 *Translated Text*

IIi9 ¶ On the 15th of Du'uzu the Arrow becomes visible, and
IIi10 4 minas is a daytime watch, 2 minas is a nighttime watch.
IIi11 The Sun which rose towards the North with the head of the Lion
IIi12 turns and keeps moving down towards the South at a rate of 40 NINDA per day.
IIi13 The days become shorter, the nights longer.

---

IIi14 ¶ On the 15th of Tešrītu the Sun rises in the Scales in the East,
IIi15 and the Moon stands in front of the Stars behind the Hired Man, 3 minas is a daytime watch, 3 minas is a nighttime watch.

---

IIi16 ¶ On the 15th of Ṭebetu, the Arrow becomes visible in the evening. 2 minas is a daytime watch, 4 minas is a nighttime watch.
IIi17 The Sun which rose towards the South with the head of the Lion turns and
IIi18 keeps coming up towards the North at a rate of 40 NINDA per day. The days become longer, the nights become shorter.

---

IIi19 ¶ On the 15th of Nisannu the Moon stands in the evening in
IIi20 the Scales in the East, and the Sun in the West in front of the Stars behind
IIi21 the Hired Man. 3 minas is a daytime watch. 3 minas is a nighttime watch.

---

IIi22 ¶ On the 15th of Nisannu, on the 15th of Du'uzu, on the 15th of Tešritu, on the 15th of Ṭebetu,
IIi23 you observe the risings of the Sun, the visibility time of the Moon, the appearances of the Arrow,
IIi24 and you will find how many days are in excess.

---

### 4.6.4 *Analysis*

#### 4.6.4.1 *Rhetorical Devices: Conclusion, Direct Address*

Section g contains a conclusion, consisting of a generalized, procedural summary statement. As in the earlier *ziqpu* star texts (subsection e-1), it directly addresses the user of the text:

> *you observe*
> *you will find how many days are in excess*

The reader is instructed to "observe" and "find" (calculate) the number of days "in excess," to check the expected risings of the stars on the evenings of the solstices and equinoxes as given; that is, to cross-check solar, lunar and stellar phenomena, so that he may "find how many days are in excess." The goal of these observations can only be to determine if, and by how many days, the non-intercalated civil calendar has fallen behind the true stellar/solar calendar (see 4.6.1 above), although the goal is not stated explicitly.

#### 4.6.4.2 *Discourse Forms: Space and Time*

Section g consists of a series of complete, complex sentences. Four entries (4.6.2) correlate the position and motion of the Sun, the Moon and selected stars at the times of the equinoxes and solstices, together with the temporal and spatial parameters on their appearance. The fifth entry, the summary statement, concludes the section.

The first four entries use absolute (see 3.1.2) coordinate systems of space (cardinal directions) and time (calendar dates and *minas*, units of the water clock). The length of a "watch," given in *minas*, differentiates the solstices, when day and night are of unequal duration, from the equinoxes, when day and night have the same duration.

The solstice entries describe path and motion in both absolute terms (*towards the South*) and intrinsic terms (*with the head of the Lion*). The rate of the Sun's movement through space is given using a fixed measure of 40 NINDA per day. The equinox entries use both absolute terms (East, West) and relative terms (*in front of the Stars, behind the Hired man*) to describe the locations of the Sun and Moon.

Time is central to the content of section g. This is reflected in the multiple ways in which time is marked. Each entry in section g begins with a fixed, absolute temporal reference, a calendar date. The solstice entries include two further types of temporal descriptions. The rate at which the Sun changes position along its North-South annual trajectory is expressed in terms of a fixed, absolute measure, forty NINDA per day (NINDA = 1/3,600 of a *mina*: base 60: 0;0,1):

> *The sun... keeps moving down toward the South at a rate of 40* NINDA *per day*

Also, both solstice entries end with a reference to time changes associated with the astronomical event, expressed in relative terms:

> *the days/nights become shorter/longer*

4.6.4.2.1 *Complexity*

The complexity of the discourse reflects the complexity of the astronomical phenomena being described. Longer, more complex sentences are used to describe the solstices than the equinoxes. At the equinoxes, the "Sun-god" simply continues his movement to the North or South, but at the solstices, the "Sun-god" must make a decision to stop, turn around, and proceed in the opposite direction, entailing a more complex description.[11]

4.6.4.2.2 *Generalized Expressions*

While the content refers to specific calendar dates, those dates and their associated stellar configurations can be expected to recur. That is, they are not simply single observations, but are rather general statements that hold for given dates:

> on *(date) (star) becomes visible*
> on *(dates) you observe*

This, of course, is also true for other sections of MUL.APIN which give dates for specific astronomical events (see e.g. 4.2.4.1.1).

### 4.6.4.3 *Dividing Lines*

The placement of dividing lines within a single unit of text appears to be motivated by the astronomical content, setting off four distinct astronomical events—two solstices and two equinoxes. However, an additional line sets off the statement that summarizes all four events. This summary statement repeats each of the four dates in the preceding

[11] For an illustration of the MUL.APIN model of solar motion, see the figure in Horowitz, 1998:173. For the Sun's role in determining the changes of the seasons in Akkadian literature see Lambert, 1960:137; 180–1 (*The Šamaš Hymn*).

text, and clearly implies that the four events are seen as related by the ancient compilers of the text. The openings of all five entries, including what we have just identified as the summary statement, begin with DIŠ. If our analysis is correct, this is the first summary statement thus far in the series marked by DIŠ.

### 4.6.5 *Categories*

In section g, the summary entry re-states the four calendar dates in sequence, followed by instructions to the reader of the text regarding the procedure to be carried out on those dates. In other words, the four dates and their related stellar configurations are grouped according to a common procedure, finding how many days are in excess.

In the same sense that summary statements in earlier sections have concluded by naming the taxonomic category of objects described in the foregoing text, this entry designates the procedure common to the foregoing entries. This entry could thus be characterized as designating a procedural "category."

## 4.7 *Sections h and i, MUL.APIN II i 25–71; plus Gap A 1–7, from Section j*

These sections share a common concern with wind and weather. Wind and weather, being phenomena of the sky, were considered to belong to the same realm as astronomy and astrology by ancient Mesopotamians, including the composers of MUL.APIN. Since the concern with weather is unique to these sections (with the minor exception of several omens in section m), we consider Hunger and Pingree sections h and i, along with Gap A, lines 1–7, as a unit. Gap A, lines 1–7, are assigned by Hunger and Pingree, in their translation, to section j. We refer to it below as j-1.

Beyond a concern with wind and weather, the structure of this portion of MUL.APIN is difficult to analyze in a coherent manner. It includes somewhat obscure references to offerings and to horses "touching bitumen" and "drinking water." There is no recognizable astronomical framework within which to approach the loosely-related sequence of topics, nor is there a coherent textual structure that would allow us to identify underlying rhetorical goals.

At best, we can identify a concern with planets and atmospheric phenomena that lead toward a rough scheme for locating the four cardinal points (see 4.7.4). This allows for inclusion of the very formal set of entries relating to the four seasons, with which the section concludes.

The following five subsections were identified by noting the placement of dividing lines on the original cuneiform tablets, together with content and thematic considerations. On our view, these divisions give the best approximation of coherence in this section of text.

### 4.7.1 *Subsection h-i-1, MUL.APIN II i 25–37*

#### 4.7.1.1 *Astronomical Content*

Subsection h-i-1 has three parallel sub-divisions, each for a different month of the year, each following the same pattern. It first gives the dates of the risings of stars, followed by instructions to observe "the wind that blows." Each sub-division then moves on to an obscure discussion of offerings, horses, bitumen, and river water.

This content is perplexing to the modern reader, although the ancient reader of MUL.APIN presumably was expected to understand what was meant.

#### 4.7.1.2 *Textual Form*

Subsection h-i-1 contains descriptions of stellar phenomena, weather (wind), and horses. Metaphor may figure here, in that the description of horses "touching bitumen" and drinking water can be construed either literally or metaphorically (see 4.7.2.5). "Guarding" can also be taken to mean watching, or more formally, to make astronomical observations.[12]

The rhetorical device of direct address is repeated and prominent. Throughout, discourse forms appropriate to the science of astronomy are interleaved with forms more characteristic of myths and omens. Thus, metaphorical, anomalous, and obscure expressions appear alongside more objective, observational statements, illustrating the transitional nature of MUL.APIN.

---

[12] All these are possible nuances of the Akkadian verb *naṣāru*.

#### 4.7.1.3 *Translated Text*

IIi25 ¶ On the 10th of Ululu the star of Eridu becomes visible, on the 15th ŠU.PA;
IIi26 on the day their stars become visible you observe their risings, their glow, and
IIi27 their...., and the wind that blows; you guard (?) the horses
IIi28 so that they do not drink water from the river.
IIi29 When their stars have been made visible,
IIi30 you present offerings to them; horses
IIi31 will touch bitumen and drink water from the river.

---

IIi32 ¶ On the 5th of Arahsamnu the Scorpion becomes visible, on the 15th the breast of the Scorpion;
IIi33 on the day they become visible you observe the wind that blows.

---

IIi34 ¶ On the 15th of Addaru the Fish [becomes visible] in the morning, in the evening the star of Eridu [becomes visible....]
IIi35 their stars....[....];
IIi36 on the day they become visible [you observe] their risings, their glow,
IIi37 their...., and the wind that blows.

---

#### 4.7.1.4 *Analysis*

##### 4.7.1.4.1 *Rhetorical Devices: Direct Address*

Subsection h-i-1 contains five (four written, one restored by Hunger and Pingree in missing text) instances of direct address of the user of the text:

*you observe*
*you guard*
*you present offerings*
*you observe*
[*you observe*]

The text refers to offerings, "guarding" of horses, to bitumen and to water, and whether these references are literal or metaphorical, the implication seems to be that the reader of the text, or "the observer of the sky," would understand the allusions. In other words, the text as written in subsection h-i-1 assumes prior knowledge, or a set of background assumptions, that would enable a reader to draw the correct inferences from the text and to understand the actions and procedures to which the text refers. Unfamiliarity with these background assumptions renders the text incomprehensible to the modern reader.

#### 4.7.1.4.2 *Discourse Forms: Space and Time*

Subsection h-i-1 uses reference to absolute (calendar dates) and relative (when the stars become visible; in the morning) coordinate systems of time, in descriptions of the parameters governing the appearance of the stars, associated winds, and events. Each line-demarcated subsection begins with a calendar date on which specific stars rise, and continues with multiple space-time references.

#### 4.7.1.4.3 *Generalizations*

While there is an absence of coherence in overall content, each line-demarcated subsection begins with a general form that states a calendar date and the rising of different stars:

*On (X date) (Y star) is observed.*

#### 4.7.1.5 *Categories*

The content and intent of subsection h-i-1 is too vague, on our reading, to warrant classification.

### 4.7.2 *Subsection h-i-2, MUL.APIN II i 38–43*

#### 4.7.2.1 *Astronomical Content*

Subsection h-i-2 describes five planets, Jupiter, Venus, Mercury, Mars, and Saturn (with alternate names for Mercury and Saturn), together with stars and winds. Like the preceding subsection h-i-1, it also includes references to horses and bitumen.

#### 4.7.2.2 *Textual Form*

Subsection h-i-2 first names planets in a format similar to the *ziqpu* star text (subsection e-1, above), that is, as a continuous complex subject rather than in a list format with discrete entries. It follows with a partially preserved (text missing on the original tablet) summary statement that refers to the planets alone, and then a summary statement concerning the stars, winds, offerings, horses and bitumen.

#### 4.7.2.3 *Translated Text*

IIi38 Jupiter, Venus, Mercury, whose name is Ninurta, Mars, Saturn,
IIi39 [also called] "the Scales" (or) "Star of the Sun."

---

IIi40 [These are the gods (?) who] keep changing their positions and their glow
IIi41 [and] touch [the stars of the sky]; on the day their stars become visible, you observe their risings, their glow,
IIi42 [their....], where they become visible, and the wind that blows: on the day they become visible,
IIi43 you present offerings to them; horses will touch bitumen.

---

#### 4.7.2.4 *Analysis*

##### 4.7.2.4.1 *Rhetorical Devices: Conclusion, Direct Address*

The last expression of the second part of subsection h-i-2 instructs the reader of the text to carry out observational and ritual tasks:

*you observe their risings*
*you present offerings*

This part of h-i-2 may form an *inclusio* with the opening part of subsection h-i-1, above, which gives the more detailed exposition of this obscure topic.

##### 4.7.2.4.2 *Discourse Forms: Space and Time*

Subsection h-i-2 names the planets and describes their visibility and movements through space and time in relative terms:

*on the day their stars become visible*
*where they become visible*
*changing their positions and their glow*

As in the earlier section, the references to offerings, horses, and bitumen are obscure to the modern reader.

##### 4.7.2.4.3 *Generalizations*

The initial expression in the second part of subsection h-i-2 is a generalization over the category of planets named in the first part of the subsection:

*these are the gods...*

Similarly, the expression beginning:

*on the day their stars become visible*

is a generalization over observations associated with the dates of the planets' appearance.

### 4.7.2.5 *Categories*

In subsection h-i-2 the planets appear in a format similar to the *ziqpu* star text (subsection e-1, above); that is, as a continuous complex subject that continues over multiple lines, here demarcated by horizontal rulings at its beginning and end. This marks the set of planets as a conceptual unit. The subsection ends with a summary statement, which Hunger and Pingree restore so as to explicitly name the set of planets (as in 4.5.2.5, above):

*these are the gods*

This restoration is plausible, but of course not certain.

The remainder of the content, below the set of planets, once again includes obscure references to horses, winds, and bitumen. "Horses," perhaps, could be taken as a metaphor for winds, as they are in Mesopotamian astro-mythological writings,[13] but what could "horses will touch bitumen" mean? Could this somehow be a short astrological

[13] Cf. following sets of allusions linking winds, animals, and deities in 'The Great Star List,' a cuneiform astrological work:

| | |
|---|---|
| In the south wind | cattle |
| In the north wind | sheep |
| In the east wind | horses |
| In the west wind | asses |

--------

| | |
|---|---|
| Anne-amaru (means) | Deluge of the Sky |
| Hursag-sa (means) | Mountain Coverer |

--------

(These are the names of) the two calves of (the storm-god) Adad

--------

| | |
|---|---|
| Gišlam-šaršar (means) | Mingler of heaven and earth |
| Balagdi-amaru (means) | Harp of the Flood |

--------

(These are the names of) the two horses of the Flood

--------

See Koch-Westenholz 1995:202–205 with previous discussion in Weidner 1959–60: 110–111.

comment of the type familiar to the editors of MUL.APIN from astronomical omens (see section m below) and related materials?[14] What does all this have to do with planets, winds, and stars? We simply cannot answer such questions on the basis of our current understanding of the text, but it seems obvious that the ancient writers or compilers of the treatise had something specific in mind.

### 4.7.3 *Subsection h-i-3, MUL.APIN II i 44–67*

#### 4.7.3.1 *Astronomical Content*

Subsection h-i-3 contains detailed planetary phenomena. Periods of visibility and invisibility of the planets are described, with their associated cardinal directions and perceptual properties, such as color and glow.

#### 4.7.3.2 *Textual Form*

Subsection h-i-3 gives detailed, elaborated, and repeated descriptions of planetary phenomena. These descriptions include multiple temporal and locative references, fixed and relative. In the extended Mercury description, discussed below, there is also some text that resembles the omen material in section m. Entries for each planet are separated by dividing lines.

#### 4.7.3.3 *Translated Text*

IIi44 ¶ Venus disappears in the East and remains (invisible) in the sky for a month,
IIi45 or for a month and 15 days, or it remains for 2 months, and
IIi46 becomes visible in the West.

---

IIi47 ¶ Venus disappears in the West and becomes visible (again) in the East on the day she disappears; or, for 3 days,
IIi48 thirdly, for seven days, fourthly, for 14 days, she remains (invisible) and then rises.

---

[14] A collection of such short comments is available in Enuma Anu Enlil Assumed Tablet 50; see Reiner & Pingree 1981:28–51; see also Horowitz and Oelsner, 1997–98:176–182.

IIi49 ¶ Jupiter disappears in the West and remains (invisible) in the sky for 20 days,
IIi50 or remains for a month, and rises and becomes visible in the East in the path of the Sun.

---

IIi51 ¶ Mars disappears in the West and remains (invisible) in the sky for 2 months, or for 3 months and 10 days, or for 6 months
IIi52 and 20 days, and rises and becomes visible in the East in the path of the Sun.

---

IIi53 ¶ Saturn disappears in the West, remains (invisible) in the sky for 20 days, and becomes visible in the path of the Sun.

---

IIi54 ¶ Mercury, whose name is Ninurta, becomes visible either in the East or in the West, and
IIi55 stands in the sky for 7 days, or for 14 days, thirdly, for 21 days, fourthly, for a month, fifthly, for a month and 15 days,
IIi56 and when it disappears it remains (invisible) for as many days as it stood in the sky
IIi57 and rises and becomes visible either in the East or in the West in the path of the Sun. (If ) this star becomes visible in winter, (there will be) rain and flood;
IIi58 (if ) it becomes visible at harvest time, you observe its glow, its…., where it becomes visible, and the wind that blows.
IIi59 This star is either red and bright, or yellow and dark.

---

IIi60 ¶ Jupiter becomes visible in th East, stands in the sky for one year, and disappears in the West.

---

IIi61 ¶ Venus becomes visible either in the East or in the West, stands in the sky for 9 months, and disappears.

---

IIi62 ¶ Mars becomes visible in the East, stands in the sky for one year and 6 months, or for one year and 10 months, or for 2 years,
IIi63 and disappears in the West. This star shows either redness and is bright, or is.… and small.

---

IIi64 ¶ Saturn, also called the Scales (or) star of the Sun, becomes visible in the East,
IIi65 stands in the sky for one year, and disappears in the West. This star is either red or white.

---

IIi66 ¶ Mercury, whose name is Ninurta, becomes visible and disappears
IIi67 within a month either in the East or in the West.

---

### 4.7.3.4 *Analysis*

#### 4.7.3.4.1 *Discourse Forms: Complexity*

Subsection h-i-3 illustrates an increase in the detail and consequent syntactic complexity of the expressions used to describe the planetary phenomena. The compilers/authors mark this complexity in the second entry for Venus, in which a series of parameters on the appearance of the planet appear with ordinal expressions: "thirdly" and "fourthly." In subsection h-i-3, the first instance of a disjunction appears in the translation, reflecting Akkadian constructions equivalent to English contrastives "or," or better, "either—or," but not at the same time.[15]

The entries are, for the most part, complete sentences, with verbs and modifiers, and multiple clausal embeddings.

#### 4.7.3.4.2 *Discourse Forms: Space and Time*

Subsection h-i-3 contains spatial reference to absolute (East, West) and relative (the sky, the path of the Sun) coordinate systems. We have seen throughout MUL.APIN that planet "location" is more accurately characterized by path or trajectory, and much of subsection h-i-3 is devoted to tracking planetary movement across the skies.

Subsection h-i-3 uses multiple references to diverse absolute intervals of time:

*for three days*
*for a month*
*for a month and 15 days*
*for one year and six months*

Notably absent from this list are absolute calendar dates, but this is to be expected, since planetary phenomena do not repeat themselves on an annual basis.

#### 4.7.3.4.3 *Generalizations*

Subsection h-i-3 is a series of generalized expressions. The recurring temporal and locative parameters governing the appearance of the planets, and in some cases, associated atmospheric conditions and ritual practices, are given.

---

[15] The "either-or" sense of the disjunction is acheived in the original Akkadian by constructs *lu* _____ *lu* _____, *ēma* _____ *ēma* _____, and symmetry: ***lu*** *ina īṣt Šamši* ***lu*** *ina ereb Šamši* **ēma** *arhi innammarma* **ēma** *arhi itabbal.*

#### 4.7.3.5 *Categories*

Subsection h-i-3 describes the trajectories of planets, and thus foregrounds movements, periods of visibility, perceptual properties, and the like. It also details the perceptual characteristics of the planets: their color, glow, and brightness. Perceptual characteristics often form part of definitions, and are one of the bases by which object identifications are made, although no concluding statement appears, and no attempt is made to define or categorize the foregoing phenomena. This subsection, like the foregoing subsections of h-i, can be seen as leading toward the final two subsections (4.7.4, 4.7.5) below.

#### 4.7.3.6 *Minor Textual Form: Description of Mercury*

In subsection h-i-3, the planet Mercury has by far the most complex text associated with it. This was observed also in section a (4.1.4.3), above, and may be because Mercury is the most difficult to observe among the heavenly bodies. However, the entry for Mercury in subsection h-i-3 is also distinguished by the intrusion of astrological material, an omen whose protasis relates the visibility of Mercury in winter, and predicts a meteorological apodosis (rain and flood). This is followed by a reference to the visibility of Mercury at harvest time being related to winds and changes in color.[16]

### 4.7.4 *Subsection h-i-4, MUL.APIN II i 68–71*

#### 4.7.4.1 *Astronomical Content*

The content of subsection h-i-4 is more specific than the text indicates. The stars are to be found in their required positions, vis-à-vis the winds, only for a short time in the spring (Horowitz, 1998:198–99). The intent of the lines is thus something like "when you observe the direction of the winds, and the following stars are located by the following winds, then, on this particular day, the stars indicate which wind is blowing."

[16] The Mercury omens section of Enuma Anu Enlil has not yet been edited with translation and notes. For omens and reports relating to Mercury see Hunger and Pingree 1999:136–138 and Koch-Westenholz 1995:127–128.

### 4.7.4.2 *Textual Form*

Subsection h-i-4 consists of four lines, one very long complex sentence that instructs the reader of the text to carry out an astronomical procedure, related to the position of the wind and stars.

### 4.7.4.3 *Translated Text*

IIi68 ¶ If you are to observe the direction of the winds: the Wagon lies across where the North wind rises,

IIi69 the Fish lies across where the South wind rises, the Scorpion lies across where the West wind rises,

IIi70 the Old Man and the Stars stand where the East wind rises.

IIi71 On the day of your observation the stars will indicate to you which wind blows.

---

### 4.7.4.4 *Analysis*

#### 4.7.4.4.1 *Rhetorical Devices: Direct Address, Procedures*

Subsection h-i-4 instructs the reader to observe the winds and stars. The opening forms an introduction that specifies the objective of the observation:

> *if you are to observe the direction of the winds*

and concludes with what the observer can expect to be the result of the procedure detailed in the text:

> *the stars will indicate to you which wind blows*

In both cases, the author addresses the observer using second person, "you."

#### 4.7.4.4.2 *Discourse Forms: Space and Time*

Subsection h-i-4 locates astronomical phenomena with respect to the four cardinal directions, however, embedded in expressions that locate the stars "across" the rising of the four winds. The four directions can be regarded as absolute (see 3.1.3), although in a limited sense, since ancient Mesopotamians conceived of North, South, East, and West as quadrants of Heaven and Earth rather than as fixed points on the compass, such as due North, etc., as in the modern conception (Horowitz, 1998:193–207; 259–260).

#### 4.7.4.4.3 *Generalizations*

Subsection h-i-4 is a generalization over conditions for observing the winds and identifying their directions. The form has been encountered earlier:

*on (X day) (Y) occurs.*

### 4.7.4.5 *Categories*

Subsection h-i-4 builds to a specific procedural statement directed to the reader of the text:

II i 71 *On the day of your observation the stars will indicate to you which wind blows.*

The concluding expression of the subsection identifies a procedure and an expected outcome, and thus can be considered a procedural category (see 4.6.5 above): how to use the star positions to identify, or categorize, the four cardinal wind directions.

## 4.7.5 *Subsection j-1, Gap A 1–7*

### 4.7.5.1 *Astronomical Content*

Subsection j-1 is a scheme relating the position of the Sun with the weather over what can only be the four seasons of the year as defined in MUL.APIN. The Sun is in the Northern part of the sky (path of the Enlil stars) during summer, the Southern part of the sky (path of the Ea stars) during winter, and in between (the path of the Anu stars) during spring and fall (Horowitz, 1998:172–174). Hence, the section is divided into four subsections by horizontal rulings.

### 4.7.5.2 *Textual Form*

Subsection j-1 presents four parallel sentences, with neither introduction nor summary statement. Each sentence begins with a conditional phrase that specifies an interval of time. Each then details the position of the Sun during the specified interval and associated weather. The structure of four parallel sentences matches the four seasons of the year.

#### 4.7.5.3 *Translated Text*

| | |
|---|---|
| IIGapA1 | ¶ From the 1st of Addaru until the 30th of Ajjaru the Sun stands in the path of the Anu stars; |
| IIGapA2 | wind and weather. |
| IIGapA3 | ¶ From the 1st of Simanu until the 30th of Abu the Sun |
| IIGapA4 | stands in the path of the Enlil stars; harvest and heat. |
| IIGapA5 | ¶ From the 1st of Ululu until the 30th of Arahsamnu the Sun |
| IIGapA6 | stands in the path of the Anu stars; wind and weather. |
| IIGapA7 | ¶ From the 1st of Kislimu until the 30th of Šabaṭu the Sun stands in the path of the Ea stars; cold. |

#### 4.7.5.4 *Analysis*

##### 4.7.5.4.1 *Discourse Forms: Space and Time*

Subsection j-1 is a series of parallel, generalized expressions that locate the Sun relative to the paths of the stars. The temporal parameters on the appearance of the Sun in these paths are expressed in absolute terms (calendar dates), thus defining the seasons. Associated weather conditions are given for each season.

##### 4.7.5.4.2 *Rhetorical Devices*

Subsection j-1 offers neither introduction nor summary statement, nor does it directly address of the user of the text.

##### 4.7.5.4.3 *Generalizations*

The form of the expressions in subsection j-1 involves intervals between dates that divide the solar year into four equal seasons:[17]

*from (X date) until (Y date); (Z conditions)*

The astronomical content is expressed in a detailed, precise, and general form. The statements reference a fixed body of knowledge, events that can be expected to occur with some reliability. In this limited sense, they can be regarded as axiomatic. This may be intentional, as the information given here serves to define both the boundaries of the

[17] These seasons begin midway between the equinoxes and solstices, rather than on the equinoxes and solstices as in our calendar.

stellar paths, and the first and last dates of the four solar seasons, by means of the location of the Sun on the dates given.

## 4.8 *Subsections j-2 and j-3, MUL.APIN II Gap A8-II ii 20*

Hunger and Pingree call their section j "the second intercalation scheme," placing it together with Gap A8. On our analysis, this section of text contains three distinct pieces of material. The first we analyzed above as subsection j-1 (4.7.5), treating it with the foregoing text because of its common concern with wind and weather. The second intercalation scheme, proper, does not begin until Gap A8.

Various rules and procedures for determining when and how to intercalate are followed by a short three-line subsection giving historical information regarding intercalary months (MUL.APIN II ii 18-II ii 20). On our analysis, then, the text is more varied than is indicated by Hunger and Pingree.

### 4.8.1 *Subsection j-2, MUL.APIN II Gap A8-II ii 17*

#### 4.8.1.1 *Astronomical Content*

Subsection j-2 consists of an intercalation scheme. It prescribes the astronomical rules and procedures for determining whether a given year is a normal lunar year, with 12 lunar months, or a leap year, with 13 months. The rules are expressed in the following manner:

If the constellation "the Stars" (the Pleiades) and Moon are in conjunction on New Year's Day (the first day of the month of Nisannu in Mesopotamian calendar) then the year is normal. But if this conjunction between the constellation "the Stars" and the moon has not occurred by the third of Nisannu, then the year is a leap year.

Thirteen rules of this type are followed by a mathematical scheme extrapolated from the observable phenomena, and a mathematical proof for the intercalary procedure.

#### 4.8.1.2 *Textual Form*

Subsection j-2 consists of thirteen parallel statements describing the procedures for judging whether a year is "normal" or a leap year.

These thirteen statements, or rules, are followed by a summary statement referring to the stars of Anu, Enlil and Ea, the three categories of stars identified by the star catalogues at the beginning of MUL. APIN in section a. The reader of the text is instructed to observe these categories of stars, to "compute," to "predict" and to "proclaim" the year to be either a regular year or a leap year in accordance with the outcome of these procedures.

The summary statement is followed by two additional sections of text. These are more abstract and generalized than earlier forms of expression found in the treatise, containing extrapolated, axiomatic mathematical expressions.

#### 4.8.1.3 *Translated Text*[18]

| | |
|---|---|
| IIGapA8 | [¶ If on the 1st of Nissanu] the Stars and the Moon are in conjunction, this year is normal. |
| IIGapA9 | [¶ If] on the 3rd [of Nissanu] the Stars and the Moon are in conjunction, this year is a leap year. |
| IIGapA10 | [¶ If] the Stars become visible [on the 1st of Ajjaru], this year is normal. |
| IIGapA11 | [¶ If] the Stars become visible on the 1st of [Simanu], this year is a leap year. |
| IIGapA12 | [¶ If on the 15th of Du'uzu] .... [....] |
| IIGapA13 | [¶ If] the Arrow becomes visible [on the 15th of Abu], this year is [a leap year.] |
| IIGapA14 | [¶ If] ŠU.PA becomes visible [on the 15th of Ululu], this year is [normal.] |
| IIGapA15 | [¶ If] ŠU.PA becomes visible [on the 15th of Tešritu], this year is [a leap year.] |
| IIii1 | [¶ If] the Stars and the Moon are in conjunction [on the 15th of Arahsamnu], this year is normal. |
| IIii2 | [¶ If] the Stars and the Moon are in conjunction on the 15th of [Kislimu], this year is a leap year. |

[18] The format here is, as in omens, i.e. if... then... In this format the single DIŠ at the start of the line serves two functions. It represents both the conditional "If" and the DIŠ marking the start of entries.

IIii3 [¶ If] the Arrow [becomes visible] in the East in the evening on the 15th of Ţebetu, [when (?) the Sun] rises towards the South, turns and keeps coming up towards the North, this year is normal.

IIii4 [¶ If] the Arrow becomes visible in the evening on the 15th of Šabaţu, this year is a leap year.

---

IIii5 [¶ If] the Fish and the Old Man become visible on the 15th of Addaru, this year is normal.

IIii6 [¶ If] the Fish and the Old Man become visible on the 15th of Nisannu, this year is a leap year.

---

IIii7 You lo[ok(?)] for the rising (?) and.... of the stars of Ea, Anu and Enlil

IIii8 and name this year;

IIii9 when...., you compute [....] and year, and

IIii10 for the third year you make a prediction, and proclaim this year a leap year.

---

IIii11 ¶ To...the day of disappearance of the Moon for 12 months, you proclaim an intercalary month in three years (variant: the third year);

IIii12 10 additional days in 12 months is the amount for one year.

---

IIii13 If you are to find the correction for day, month, and year:

IIii14 you multiply 1, 40, the correction for a day, by one month,

IIii15 and you find 50, the correction for one month; you multiply 50, the correction for one month,

IIii16 by 12 months, and you find 10 additional days, the amount for one year.

IIii17 In three years (variant: the third year) you proclaim (this year) a leap year.

---

#### 4.8.1.4 *Analysis*

##### 4.8.1.4.1 *Discourse Forms: Time and Space*

Subsection j-2 uses references both to absolute (calendar dates) temporal and relative (conjunction of the Moon and constellations) spatial-temporal coordinate systems. A more detailed locative expression is found in the entry for "the Arrow" in the month of Ṭebētu (II ii 3) which parallels the more complex motion of the Sun at the winter solstice.

##### 4.8.1.4.2 *Rhetorical Devices: Summary Statement, Direct Address*

Subsection j-2 concludes with three successive segments of text in which the reader of the text is addressed. These are II ii 7–10, II ii 11–12, and II ii 13–17. Horizontal rulings separate the three, but only the second is marked by DIŠ. The reason for this is not certain, but we find a parallel phenomenon in the summary to section k, below.

The first summary statement refers to the observations and calculations the reader of the text is to perform in order to achieve the goal of the procedure, whether the year is a leap year or a normal year:

II ii 7–8 *You lo[ok (?)] for the risings (?) and... of the stars of Ea, Anu and Enlil/..... and name this year/*

The reader is then instructed to "compute," "predict," and "proclaim," based on the results of the astronomical procedures referred to immediately prior:

II ii 9–10 *when you compute ... and year, and/for the third year you make a prediction, and proclaim this year a leap year*

The second and third summary statements are discussed in 4.8.1.4.4 below.

4.8.1.4.3 *Generalizations: Decision Rules Expressed as Conditionals*

Subsection j-2 begins with thirteen entries giving the procedures, or decision rules, for assigning years to types, or categories, in the form of generalized, conditional expressions beginning with "if." The astronomer decides whether a year is a leap year or a normal year based on these rules. The lines in the text separate out different forms of expression, or different types of rules. The entries are in pairs and have the form:

*if, on (date) (expected observation), (then) year (normal)*
*if, on (date) (expected observation plus one month), (then) year (leap)*

4.8.1.4.4 *Rhetorical Devices: Mathematical Procedure*

The next segment opens with a short passage which instructs the reader to intercalate the calendar once every third year on the basis of a given mathematical-calendrical principle:

II ii 11 *To... the day of disappearance of the Moon for 12 months, you proclaim an intercalary month in three years (variant: the third year)*

This is followed by a statement in a general, or axiomatic, form:

II ii 12 *10 additional days in 12 months is the amount for one year.*[19]

[19] These lines refer to a 364-day year consisting of 12 lunar months (= 354 days) and 10 additional days per year, with the difference of 30 days every third year

Subsection j-2 concludes with a review of the mathematical procedure. The calculation of the number of additional days per 12 months (10 additional days, according to the preceding section) is formalized (II ii 13–17):

| | |
|---|---|
| II ii 13 | If you are to find the correction for day, month, and year: |
| II ii 14 | you multiply 1, 40, the correction for a day, by one month, |
| II ii 15 | and you find 50, the correction for one month; you multiply 50, the correction for one month, |
| II ii 16 | by 12 months, and you find 10 additional days, the amount for one year. |
| II ii 17 | In three years (variant: the third year) you proclaim (this year) a leap year. |

4.8.1.5 *Categories*

Subsection j-2 consists of procedures for categorizing years as "normal" or "leap." The process of categorization finds expression twice, once as "naming" (the Akkadian verb *nabû*, II ii 8) and once as "proclaiming" (the Akkadian verb *qabû*, II ii 10). This goal is realized by means of the thirteen decision rules, and then explained in more general terms in the three concluding segments just discussed. The thirteen decision rules are followed by a now partially broken summary statement that refers to the three main categories of stars in the MUL.APIN system:

... *the stars of Ea, Anu, and Enlil*

Thus, in this summary, we find these three categories of stars appearing in a single expression, most likely to be restored *the risings and settings of the stars of Ea, Anu, and Enlil*, in which these three groups of stars form a conceptual unit (i.e. probably all the stars in the sky). This stands in marked contrast to MUL.APIN section a where we found the individual stars of the Paths of Anu, Enlil, and Ea listed as three separate lists or categories, each with their own summary line.

The mathematical scheme that follows (4.8.1.4.3 above) specifies how an intercalary month is to be determined, or categorized, and

accounted for by the addition of a leap month. This sequence of two 12-month regular years and one 13-month leap year yields the 37 month cycle which was known in Ancient Mesopotamia as early as the late 4th millennium (Nissen, Damerow, and Englund 1993:36).

defines the "amount for one year," that is, the number of days that determine the difference in the calculation.

The mathematical proof that concludes subsection j-2 gives a procedural definition (see 4.6.5, above) of the "correction," for day, month, and year. This astronomical correction is the ultimate objective of the intercalation scheme. Each of these three concluding segments can be considered as procedural definitions over categories of years, amounts, and corrections respectively.

### 4.8.2 *Subsection j-3, MUL.APIN II ii 18–20*

#### 4.8.2.1 *Content and Analysis*

The very short subsection j-3 appears to provide historical information regarding intercalary months. Three different intercalary months are identified with three different historical periods: The first, a second Nisannu (month $I_2$) during the reign of Shulgi of Ur (2096–2048); the second, a second Adaru (month $XII_2$) during the first half of the second millennium (the time of the Amorite dynasties, including Hammurabi's Babylon); and the third, a second Ulūlu (month $VI_2$) during the Kassite dynasty, which ruled over Babylonia for much of the second half of the second millennium BC (Sommerfeld, 1995).

However, there is no historical evidence to justify these three statements. They seem rather to be of a type commonly found in Mesopotamian astronomical/astrological/mystical explanatory texts that give traditionally accepted associations among dates, units of time, names of countries, parts of the universe, natural phenomena, ruling dynasties, and divinities.[20]

#### 4.8.2.2 *Translated Text*

IIii18 ¶ An intercalary Nisannu (belongs to) the reign of Šulgi;
IIii19 ¶ an intercalary Addaru (belongs to) the reign of Amurru;
IIii20 ¶ an intercalary Ululu (belongs to) the reign of the Kassites.

---

[20] For example the materials in 'The Great Star List', see Koch-Westenholz, 1995; see e.g. this chapter, footnote 13.

## 4.9 *Section k, MUL.APIN II ii 21–42*

### 4.9.1 *Astronomical Content*

Section k specifies the lengths of day and night, (in terms of the water-clock) at the solstices and equinoxes, and then gives the relevant values for the gnomon (i.e. the length of the shadow cast by the gnomon on these dates).[21]

### 4.9.2 *Textual Form*

Section k is a series of entries with parallel structure presented in a schematic form. Four subsections are delineated by lines on the original tablets. Each begins with a sentence that specifies a date together with the length of day and night using a fixed temporal reference, i.e. *minas* on the water clock, followed by a number of entries listing cubits of shadow corresponding to units of *bēru*, UŠ, and NINDA. The section is remarkable for its level of detail, including compound numbers and fractions.

The fifth subsection of section k consists of a summary statement, giving a procedure for finding the "difference" for 1 cubit of shadow.

### 4.9.3 *Translated Text*

IIii21 ¶ On the 15th of Nisannu, 3 minas is a daytime watch, 3 minas is a nighttime watch.

| | | |
|---|---|---|
| IIii22 | 1 cubit of shadow | 2½ bēru of daytime |
| IIii23 | 2 cubits of shadow | 1 bēru 7 UŠ 30 NINDA daytime |
| IIii24 | 3 cubits of shadow | ⅔ bēru 5 UŠ daytime |

---

IIii25 ¶ On the 15th of Du'uzu, 4 minas is a daytime watch, 2 minas is a nighttime watch.

| | | |
|---|---|---|
| IIii26 | 1 cubit of shadow | 2 bēru daytime |
| | 2 cubits of shadow | [1] bēru daytime |
| IIii27 | 3 cubits of shadow | ⅔ bēru daytime |
| | 4 cubits of shadow | ½ bēru daytime |
| IIii28 | 5 cubits of shadow | 12 UŠ daytime |
| | 6 cubits of shadow | 10 UŠ daytime |

21 For this section of MUL.APIN see Hunger and Pingree 1999:79–83. For the water clock and gnomon see Brown, Fermor, and Walker, 1999.

IIii29 8 cubits of shadow 7 UŠ 30 NINDA daytime
IIii30 9 cubits of shadow 6 UŠ 40 NINDA daytime
10 cubits of shadow 6 UŠ daytime

---

IIii31 ¶ On the 15th of Tešritu, 3 minas is a daytime watch, 3 minas is a nighttime watch.
IIii32 1 cubit of shadow 2 ½ bēru daytime
IIii33 2 cubits of shadow 1 bēru 7 UŠ 30 NINDA daytime
IIii34 3 cubits of shadow ⅔ bēru 5 UŠ daytime

---

IIii35 ¶ On the 15th of Ṭebetu, 2 minas is a daytime watch, 4 minas is a nighttime watch.
IIii36 1 cubit of shadow 3 bēru daytime
2 cubits of shadow 1 ½ bēru daytime
IIii37 3 cubits of shadow 1 bēru daytime
4 cubits of shadow ⅔ bēru 2 UŠ 30 NINDA daytime
IIii38 5 cubits of shadow 18 UŠ daytime
6 cubits of shadow ½ bēru daytime
IIii39 8 cubits of shadow 11 UŠ 15 NINDA daytime
9 cubits of shadow 10 UŠ daytime
IIii40 10 cubits of shadow 9 UŠ daytime

---

IIii41 If you are to find the difference for 1 cubit of shadow, you multiply 40, the difference for daytime
IIii42 and nighttime, by 7, 30, and you find 5, the difference for 1 cubit of shadow.

---

### 4.9.4 *Analysis*

#### 4.9.4.1 *Rhetorical Device: Table-Like Format*

Section k has a schematic structure. To some extent, it appears reminiscent of the list structures encountered in section a. The difference is that the section k entries are not observed natural kinds, but rather stipulated kinds. The content consists of a series of measures of time and distance, quantities derived from or inferred on the basis of man-made devices. Here there are statements of equivalence, units on the left corresponding to units on the right. Section k thus articulates, in a highly specific manner, relations between diverse absolute systems of measurement, and presents them in a systematic, table-like format.

The overall structure of the section is revealed by the editors' use of horizontal rulings and DIŠ. The section consists of four subsections with the values for one of the equinoxes and solstices in each section in the sequence spring, summer, autumn, winter. Each subsection begins with DIŠ, and closes with a horizontal ruling. The section for winter is followed by a summary statement with direct address of the reader of the text in second person:

II ii 41 *If you are to find…*

This final subsection is not marked by DIŠ. We would explain this phenomenon as follows: The four subsections for the seasons are in effect four individual entries, all marked by DIŠ, while the summary statement is not considered to be an entry. This is the same pattern that we found above in sections a and e. This analysis may also be relevant to subsection j-2 where there are three summary-type statements; two of which are similar (II ii 7–10, II ii 13–17), and one that is not (II ii 11–12). Interestingly, the two which are similar, like the summary statement in k, are unmarked by DIŠ, while the other dissimilar entry is marked by DIŠ.[22]

#### 4.9.4.2 *Rhetorical Devices: Direct Address, Summary Statement*

Section k concludes with a summary statement addressing the user of the text, in second person. The summary statement specifies the calculation to be performed using the information in the foregoing table. It is noteworthy that only the summary statement does not open with DIŠ, while the sections giving the values do.

### 4.9.5 *Categories*

The summary statement of section k, insofar as it specifies not only a procedure, but also its outcome, it can be considered a procedural definition over the category "difference for one cubit of shadow" (compare with 4.6.5, 4.7.4.5, above).

Section k presents a series of stipulated, defined measurement categories through statements of equivalence, organized in the form of a table. The procedural definition for the calculation of the summary category, the "difference for one cubit of shadow," concludes the section.

---

[22] On this basis, we may speculate that summary statements without DIŠ have a different status than those marked with DIŠ from the perspective of the compilers of the text. Perhaps, those without DIŠ were added later to the series after the sequence of entries (i.e. units marked by DIŠ) had already been established. See chapter 6, 6.3.3, for further discussion of DIŠ.

## 4.10 *Section L, MUL.APIN II ii 43–II iii 15*

### 4.10.1 *Astronomical Content*

Section L is the final astronomical-mathematical section of the MUL. APIN treatise. It gives measurements for the length of day and night, for each month during the year. As in section k, this is achieved using a system of diverse units of measurement. Section L gives "ideal" intervals between sunset and moonset for new moons (the evening of the first day of a lunar month), and between sunset and moonrise on the evening of the 15th of the lunar month; this marking the beginning of the second half of the lunar month, around the time of the full moon.[23]

These two phenomena, which involve the monthly pattern of the sun and moon, are treated as a single set of data, in which there is a constant rate of change of 40 NINDA per half-month.

Section L closes with a three line summary statement, giving two procedural statements, the first of which is the procedure for determining the visibility of the moon on a given date, using the equinox as an example. The second procedure, derived from the first, gives the daily increment of change in the length of day and night.

### 4.10.2 *Textual Form*

Section L is a succession of sentences in parallel form. Each line begins with a calendar date and gives the length of nights ("watch"), and related values for intervals between solar and lunar observations. The data are reported in a more exact manner than in earlier sections of the treatise. Dividing lines are used within the section to mark off two-line entries for each month, one for the first day of the month and a second for the 15th day of the same month. Twelve months are presented in this manner. This final astronomical mathematical section of the MUL.APIN treatise concludes with a formally-expressed axiom, in the opening line of its summary statement.

---

[23] Full moons occur on the 14th or 15th of the lunar month, and are ideally marked by the full moon rising on the eastern horizon just as the sun is setting on the western horizon. The measure here of a time interval between sunset and moonrise thus refers to observations of the moon just after it is full, effectively marking the start of the second half of the month.

### 4.10.3 *Translated Text*

IIii43 ¶ On the 1st of Nisannu a nighttime watch is 3 minas 10 shekels; 12 UŠ 40 NINDA setting of the Moon.

IIii44 ¶ On the 15th of Nisannu a nighttime watch is 3 minas; 12 UŠ rising of the Moon.

---

IIii45 ¶ On the 1st of Ajjaru a nighttime watch is 2 5/6 minas; 11 UŠ 20 NINDA setting of the Moon.

IIii46 ¶ On the 15th of Ajjaru a nighttime watch is 2 2/3 minas; 10 UŠ 40 NINDA rising of the Moon.

---

IIii47 ¶ On the 1st of Simanu a nighttime watch is 2 1/2 minas; 10 UŠ setting of the Moon.

IIii48 ¶ On the 15th of Simanu a nighttime watch is 2 1/3 minas; 9 UŠ 20 NINDA rising of the Moon.

---

IIii49 ¶ On the 1st of Du'uzu a nightime watch is 2 minas 10 shekels; 8 UŠ 40 NINDA setting of the Moon.

IIii50 ¶ On the 15th of Du'uzu a nighttime watch is 2 minas; 8 UŠ [rising of the Moon].

---

IIii51 ¶ On the 1st of Abu a nighttime watch is 2 minas 10 shekels; [8 UŠ 40 NINDA setting of the Moon].

IIii52 ¶ On the 15th of Abu a nighttime watch is 2 1/3 minas; 9 U[Š 20 NINDA rising of the Moon].

---

IIii53 ¶ On the 1st of Ululu a nighttime watch is 2 1/2 minas; 10 UŠ setting [of the Moon].

IIii54 ¶ On the 15th of Ululu a nighttime watch is 2 2/3 minas; 10 UŠ 40 NINDA rising [of the Moon].

---

IIiii1 ¶ On the 1st of Tešritu a nighttime watch is 2 5/6 minas; 11 UŠ 20 NINDA setting [of the Moon].

IIiii2 ¶ On the 15th of Tešritu a nighttime watch is 3 minas; 12 UŠ rising [of the Moon].

---

IIiii3 ¶ On the 1st of Arahsamnu a nightime watch is 3 minas 10 shekels; 12 UŠ 40 NINDA setting of [the Moon].

IIiii4 ¶ On the 15th of Arahsamnu a nighttime watch is 3 1/3 minas; 13 UŠ 20 NINDA rising of the Moon.

---

IIiii5 ¶ On the 1st of Kislimu a nighttime watch is 3 1/2 minas; 14 UŠ setting of the Moon.

IIiii6 ¶ On the 15th of Kislimu a nightime watch is 3 2/3 minas; 14 UŠ 40 NINDA rising of the Moon.

---

IIiii7 ¶ On the 1st of Ţebetu a nighttime watch is 3 5/6 minas; 15 UŠ 20 NINDA setting of the Moon.

IIiii8 ¶ On the 15th of Ṭebetu a nighttime watch is 4 minas; 16 UŠ rising of the Moon.

IIiii9 ¶ On the 1st of Šabaṭu a nighttime watch is 3 5/6 minas; 15 UŠ 20 NINDA setting of the Moon.
IIiii10 ¶ On the 15th of Šabaṭu a nighttime watch is 3 ⅔ minas; 14 UŠ 40 NINDA rising of the Moon.

IIiii11 ¶ On the 1st of Addaru a nighttime watch is 3 ½ minas; 14 UŠ setting of the Moon.
IIiii12 ¶ On the 15th of Addaru a nighttime watch is 3 ⅓ minas; 13 UŠ 20 NINDA rising of the Moon.

IIiii13 4 is the coefficient for the visibility of the Moon; you multiply 3 minas, a nighttime watch,
IIiii14 by 4, and you find 12, the visibility of the Moon.
IIiii15 You multiply 40 NINDA, the difference for daytime and nighttime, by 4, and you find 2, 40, the difference of the visibility.

### 4.10.4 *Analysis*

#### 4.10.4.1 *Discourse Forms: Time and Space*

Section L is a series of parallel sentences. The sentences include fixed temporal reference to calendar dates, and give precise information regarding time intervals between sunset and moonset (at the start of the lunar month) and sunset and moonrise (at the middle of the month). Section L is distinguished from earlier sections by the detail and precision of the measurements given.

#### 4.10.4.2 *Rhetorical Devices: Direct Address, Conclusion, Axiom*

Section L concludes with a summary statement that comprises a procedural definition over the category "the coefficient for the visibility of the Moon." It directly addresses the user of the text, instructing the astronomer to carry out the procedures typical of this portion of Tablet II of the treatise. The summary statement is distinguished from those of other sections, in that it opens with a formally-expressed axiom:

II iii 13 4 is the coefficient (*igigubbu*)[24] for the visibility of the moon

[24] The term *igigubbu* (from Sumerian igi gub, literally "that which stands in front") refers to fixed standard numbers, what have been called constants or coefficients, that

This is followed by specific instructions for the calculation by which the named coefficient is to be derived. The final astronomical-mathematical section of the MUL.APIN treatise thus concludes with the most purely abstract form of expression in the entire treatise.

Again we find the phenomena we identified in Section k: entries marked by DIŠ, and the summary section with the axiom unmarked. This pattern is also found in Sections a and e.

### 4.10.5 *Categories*

Section L contains the prototype example of a stipulated category, that is, a category stipulated by verbal definition (see 3.5.1). The coefficient for the visibility of the moon is given both a numerical value and a procedural definition. The twenty-four parallel expressions that precede the axiom use statements of equivalence to define the periods of time to be used in the calculations. The summary statement is an extrapolation expressed in general terms.

## 4.11 *Section m, MUL.APIN II iii 16–iv 12*

### 4.11.1 *Content*

Section m presents astrological material. It consists primarily of omens of the type found in the Enuma Anu Enlil series.[25] For the

express the relationship between two sets of numbers in which the numbers of Set I can be multiplied by a constant (the *igigubbu*) to reach the numbers in Set II. For example, in our world, currency exchange in which X dollars multiplied by the current exchange rate = Y euros; X hours of work multiplied by the minimum wage = Y dollars in payment; or X, the length of the radius of a circle, multiplied by the constant $2\pi$ = the length of circumference of the circle. In Ancient Mesopotamia, *igigubbu* were assigned to many aspects of everyday life as well as to selected mathematical formulations and geometric shapes. An example from geometry is the *igigubbu* of the circle, the number 6, expressing the axiomatic relationship in Babylonian mathematics between the length of the radius of a circle and its circumference, which we would express as an equation: $6r = c$ (an approximation of the true value $2\pi r = c$, as the Babylonian mathematicians did not know $\pi$). A recent study of *igigubbu* in geometry, other branches of mathematics, and other aspects of life in Mesopotamia can be found in Robson 1998.

[25] For an overview of the series see Hunger and Pingree, 1999:12–20. Editions of omens from Enuma Anu Enlil include Reiner and Pingree, 1981; van Soldt 1995; Reiner, 1998; and Verderame, 2002.

Mesopotamian scribe, what we call "astrology" was a fully legitimate branch of the study of observed phenomena overhead, which explained the interrelationships between stellar, lunar, solar, and atmospheric phenomena, and the realm of human activity or experience. For the modern reader, astrology and mathematical astronomy are two separate, even opposite, fields of endeavor. However, the inclusion of astrological material in this treatise would not have seemed contradictory to the compilers or readers of MUL.APIN.

### 4.11.2 *Textual Form*

The form of section m is a series of conditional expressions (omens with a protasis and apodosis) that describe events that can be expected to transpire under certain stellar, lunar, or solar conditions. The text is divided by lines into thematic groupings not always clear to the modern reader. The subsections marked off are sometimes a group of expressions relating to a specific planet or star or stars, a particular configuration of astronomical occurrences or a particular set of outcomes, such as the rule of the king.

No introduction or summary statement is given.

### 4.11.3 *Translated Text*

IIiii16 ¶ If the light (?) of the Stars is red (?): the irrigated land will prosper.
IIiii17 ¶ If the constellation of Ištar.... s in Addaru:....
IIiii18 ¶ If one star in the constellation of Ištar is very bright: the enemy.
IIiii19 ¶ If.... 4 or 2 big ones are yellow: deaths (?).
IIiii20 ¶ If all.... are very red: flood and rain.
IIiii21 ¶ If the Fish in [....] either shows redness and is bright, or is.... and small:....

---

IIiii22 ¶ If the U.RI.RI-star becomes visible: rain and flood.
IIiii23 ¶ If the U.RI.RI-star approaches the Chariot: horses will die.
IIiii24 ¶ If the U.RI.RI-star approaches the Furrow: the same.
IIiii25 ¶ If the U.RI.RI-star approaches the Crab: .... will die.
IIiii26 ¶ If the U.RI.RI-star approaches the Crab on the right side:.... will die.
IIiii27 ¶ If the U.RI.RI-star approaches the Crab on the left side:.... will die.

---

IIiii28 ¶ If the Raven....s below to the direction of the South wind: sesame will prosper.
IIiii29 ¶ If the Raven...s above to the direction of the North wind: the barley crop will not prosper.

---

IIiii30 ¶ If the stars of the Lion....: the king will be victorious wherever he goes. This star
IIiii31 is a campaign star.
IIiii32 ¶ If the KAL.NE-star....s and approaches the 4 stars of the Stars: the ruler (?) of the city will become good (?).
IIiii33 ¶ If Jupiter is bright: rain and flood.
¶ The name of the Rainbow is "day (?) of abundance";
IIiii34 ¶ In the South, rain; in the North, flood; in the East, rain; in the West, devastation.

---

IIiii35 ¶ On the day the Lisi-star becomes visible, a man should wake up at night all that is around his house,
IIiii36 people, cattle, sheep, donkeys,
IIiii37 and he must not sleep; he should pray to the Lisi-god, then he and all that is around his house will experience success.

---

IIiii38 ¶ If in Kislimu, Ṭebetu, or Šabaṭu the left horn of the Moon is pointed
IIiii39 and looks towards the earth:....

---

IIiii40 ¶ If the Sun rises in a *nīdu*-cloud: the king will become furious and raise weapons.
IIiii41 ¶ If the sun sets in a *nīdu*-cloud: the king will die.

---

IIiii42 ¶ If a star flares up from the West and enters in the Lisi-star: there will be revolution.
IIiii43 ¶ If a star flares up from the West and enters the Yoke: there will be revolution.
IIiii44 ¶ If a star flares up from the West and enters the Moon: there will be revolution;
IIiii45 [¶ If] this star comes out (from the Moon) as three stars: unsuccessful attack.

---

IIiii46 If a star flashes from the East towards the South, passes EN.TE. NA.BAR.HUM
IIiii47 and sets in the West: for three years the land will see abundance.
IIiii48 [¶ If a star] passes from the West to the East: for three years
IIiii49 the land will experience evil.
IIiii50 ¶ If a star flares up from the middle of the sky and sets in the West: a heavy loss
IIiii51 will occur in the land.
IIiii52 [¶ If...] the Scorpion becomes visible and the South wind blows: this year will be good.

---

| | | |
|---|---|---|
| IIiii53 | [¶ | If....]....: the king of the West will prosper. |

---

| | | |
|---|---|---|
| Gap B 1 | [¶ | If] the star [of Mar]duk becomes visible at the beginning of the year: in this year the crop will prosper. |
| Gap B 2 | [¶ | If] the star of Marduk reaches the Stars: in this year the Storm god will devastate. |
| Gap B 3 | ¶ | If the star of Marduk reaches the Raven: early sesame will prosper. |
| Gap B 4 | ¶ | If the star of Marduk sees the body of a man: epilepsy (?) will seize him. |
| Gap B 5 | ¶ | If a man takes a bath in front of the star of Marduk: there will be guilt. |
| Gap B 6 | ¶ | If the star of Marduk is dark when it becomes visible: in this year there will be *asakku*-disease. |

---

| | | |
|---|---|---|
| Gap B 7 | ¶ | If the Yoke is dim when it comes out: the late flood will come. |
| Gap B 8 | ¶ | If the Yoke keeps flaring up when it comes out: the flood will be early. |
| IIiv1 | ¶ | If the Yoke keeps flaring up like fire when it comes out: the crop will prosper. |
| IIiv2 | ¶ | If the Yoke is very low and dim when it comes out: there will be no flood. |
| IIiv3 | ¶ | If the Yoke is turned towards sunset when it comes out, (if) the West wind blows and |
| IIiv4 | | turns to South: on the 10th of Ulūlu there will be destruction of the land. |
| IIiv5 | ¶ | If the Yoke is turned towards sunrise (?) when it comes out and faces the front of the sky, |
| IIiv6 | | and no wind blows: there will be famine, the dynasty will disappear; |
| IIiv7 | | omen of Ibbi-Sin, king of Ur, who went in fetters |
| IIiv8 | | to Anšan; after him his people weep (variant: fall). |

---

| | | |
|---|---|---|
| IIiv9 | ¶ | If a man is made a ruler, and the South wind blows: this man will become good. |
| IIiv10 | ¶ | If a man is made a ruler, and the North wind blows: he will eat thin (?) bread. |
| IIiv11 | ¶ | If a man is made a ruler, and the East wind blows: his days will be short. |
| IIiv12 | ¶ | If a man is made a ruler, and the West wind blows: he will not prosper. |

### 4.11.4 *Analysis*

#### 4.11.4.1 *Rhetorical Devices: Omens*

Section m represents the kinds of predictions astronomers of the time might be expected to supply a ruler or another "consumer" of their knowledge (Hunger, 1992). The omens are drawn from the tradition of the Enuma Anu Enlil series, with the occasional commentary to the omens (such as that in II iii 33) added immediately after the omen in question. Such mixing of omens and commentary is typical of surviving exemplars of many of the tablets of Enuma Anu Enlil (see e.g. Reiner & Pingree, 1981; Reiner 1998, van Soldt, 1995). As in Enuma Anu Enlil, the use of DIŠ and horizontal rulings is not always consistent or understandable to the modern reader. For example, in the aforementioned II iii 33 we have the unusual situation of two DIŠ signs on a single line of text; one marking the start of a short omen occupying only the first half of the line, and the second marking the start of the short commentary: The name of the Rainbow is "day(?) of abundance."

The kinds of predictions made in the omens, to the mind of the modern reader, lack a scientific base, to say the least. But this is not to say that the ancient astronomers would have considered the astronomical information in the previous sections to be of greater worth than this "astrological" material.

With the possible exception of a relation between seasons and weather patterns, explored in sections h and i (4.7), there seems to be little connection between the astronomical-mathematical body of knowledge and procedures that comprise the earlier sections of the treatise and section m.

Yet the omens do have "astronomical" content. They refer to stars, planets, their movements, and configurations. In fact, embedded in the omens are two intrusions. First, MUL.APIN II iii 33-4 offers a short discussion of rainbows. Then, MUL.APIN II iii 35-8 offer instructions for a ritual to be performed on the eve of the rising of the Lisi star.

The presence of the omens and related material at the end of MUL.APIN reminds us that MUL.APIN is an omnibus collection of astronomical texts. The inclusion of the omens could be based on considerations of historical completeness as easily as on considerations of usefulness or relevance, or it could reflect the history of source material.

Section m could be a collected compendium of received "knowledge" from a historical source, without which the treatise would not have been deemed complete.

Secondly, the astronomers of the day would not be immune from the question leveled at scientists throughout the ages. When rulers or ordinary folk saw the scribes spending hours, days, even lifetimes gazing at the heavens and writing down their observations and calculations, they would undoubtedly have asked why. One can imagine the Mesopotamian astronomer being asked to justify the practical value of their work.

A Mesopotamian scribe, no less than a modern scientist, would have been dependent for his living on funds dispensed from a variety of sources. Contemporary reports of astronomers to the Assyrian court (Hunger, 1992) make clear the application of astronomical knowledge to the decision making process of the Assyrian royal administration. The king wanted to know what the stars said about the future. Thus, the "applied astronomy" presented in section m must be viewed in light of the full range of the ancient Mesopotamian scribe's professional activities.

# CHAPTER FIVE

# SUMMARY OF RESULTS

The analysis in Chapter 4 details the basic features of the MUL.APIN treatise and describes how these features manifest through the progression of component sections. We have attempted to keep the analysis neutral enough to accommodate any interpretative approach, but certain patterns in the text were remarkable to us. We summarize these patterns below, as early, middle, and late, according to their relative placement within the sequence of component text sections.

## 5.1 *The Language of Space and Time*

All three frames of temporal-spatial reference—intrinsic, relative and absolute—appear throughout the component text sections of MUL. APIN. However there is a clear progression in the relative dominance of frame of reference. At the beginning of the treatise, the language of space and time appears in primarily intrinsic and relative frames of reference, while toward the end of the treatise absolute frames of reference are most in evidence.

*Early*

At the beginning of the treatise (section a) the stars are located in relation to one another by means of an intrinsic frame of reference and features of individual heavenly stars, planets, or constellations:

I i 10 *in the tail of the Lion*
I i 16 *in the cart-pole of the Wagon*

or relative frame of reference, implying but not stating a point of view:

I i 42 *behind the Field*

There is a paucity of explicit, objective locative and temporal information throughout the star lists in section a. This could be characterized as under-marking. That is, less information than seems needed to locate the stars in the sky is included.

The only exception to this early pattern is in the minor textual form noted for the planet Mercury (4.1.4.3), where we find absolute reference to the cardinal directions East and West.

*Middle*

The intermediate component text sections, in contrast to the earlier sections, contain multiple-marking of spatial-temporal parameters. This is clearest in the second *ziqpu* star text, section e-2, where five locative entries (including seven terms in total: West, East, right, left, South, middle, "opposite your breast") can be found in the first lined section of the component text. These entries include both absolute and relative locative frames of reference (4.4.2.4.3); see also 3.1.2):

I iv 10–13 *West to your right*
*East to your left*
*Your face directed towards South*
*In the middle of the sky*
*Opposite your breast*

These occur together with absolute temporal reference in the form of calendar dates that appear repeatedly throughout the component text section. The increased level of detail in spatial-temporal reference is also seen in the explicit marking of a deictic center from which the frame of reference is to be understood:

I iv 14 *opposite your breast*

referring back to the putative astronomer, or user of the text:

I iv 2 *the observer of the sky*

In each of the subsequent observations, the location of the stars is marked twice relative to the observer:

I iv 15–16 *in the middle of the sky*
*opposite your breast*

The pattern of spatial-temporal expressions in this component text could thus be characterized as over-marking, or redundant. For example, if an observer faces South, the relative locations of East and West can be inferred. They don't really need to be stated.

*Late*

In section L, the final astronomical-mathematical section of the treatise, all temporal and spatial expressions are absolute, using objective terms encountered earlier in the text: *minas*, UŠ and NINDA. *Minas* are expressed as compound numbers employing fractions, or alternately, adding the quantitative sub-unit, *shekel*, a heightened precision that first appears in the immediately preceding section k. The summary statement of section L includes the most formal and precise expression in the entire treatise, the axiom for the coefficient of visibility of the moon.

The pattern of spatial-temporal expressions in section L can be characterized as appropriately marked; that is, detailed and precise without being redundant.

We note again that that these observed patterns in the development of the language of space and time, from "early" to "late" component texts, do not represent a monotonic developmental progression through the treatise. There are exceptions to the pattern. The generalizations in the intermediate section (Chapter 4, 4.3), which appear relatively early in the text, include specific, appropriately marked, absolute expressions of degree (NINDA); also some later component sections (see 4.6.4.2) make use of relative frames of reference.

However, the broad pattern observed from early to late seems to hold. The difference between the language of space and time found in the early section a (4.1.4.2) and that found in the late section L (4.10.4.1) is quite striking. Early component sections are under-marked, middle sections are over-marked, and late sections are appropriately and precisely marked.

## 5.2 *Rhetorical Features: Introductions and Conclusions*

*Early*

The early sections of the treatise include neither introductions nor conclusions, nor any explicit marking of how the text is to be used. The summary statements in section a (4.1.5) name the categories of the preceding lists of stars, and in this sense "conclude" the sections. However, they are names only. While conceptually they function to identify the foregoing category of stars, rhetorically they seem to function as the title of the list. In a modern text, they might appear at the beginning rather than the end of the list.

*Middle*

Introductions and conclusions begin to appear with the first *ziqpu* star text, e-1 (4.4.1.4.1). The introduction to section e-1 states the goal of the procedure—observing the *ziqpu* stars—together with relevant temporal and locative information. The goal is restated in the conclusion. The central entry of the subsection consists of a continuous expression in which the names of the stars to be observed are given (4.4.1.5).

The structure of subsection e-1 as a whole can thus be viewed as identifying a single procedure. The introduction identifies the observation to be carried out, or in other words, the topic of the text (the *ziqpu* stars); the central entry lists the stars; and the conclusion restates the observational goal, and explicitly names the category of stars ("All these are the *ziqpu* stars.").

*Late*

Subsection j-2, the second intercalation scheme (4.8.1) is comprised of a series of parallel entries and concludes with a specific, general statement of the goal of the observations and calculations specified in the preceding entries. Each lined entry within this component text can be taken as specifying a procedure, or decision rule (4.8.1.4.3). Each entry, in this limited sense, conveys a content that is comparable in complexity to the entire subsection e-1: topic and goal of the procedure.

Similarly, the final astronomical-mathematical section of the treatise, section L, is a comprehensive series of measurements for the length of day and night for each month of the year (4.10.1), and the conclusion specifies in detail the relevant calculations to be carried out. As in subsection j-2, the procedures outlined in section L are more comprehensive, specific, and complex than those in earlier component sections. The concluding portions of section L include a formal proof and an axiom, detailed statements of the use to which observations in the component texts are to be put.

In comparison to the earlier components of the treatise, both j-2 and L are notable for their complexity, the number of observations and procedures referred to, the level of detail and precision of the measures and observations, and the explicitness of the procedures described.

The early component sections thus give only the category name or "title" of the list in a concluding line; in the middle sections, introductory and concluding statements of goals and procedures appear; and in the late sections, detailed, multiple statements of goals and procedures are found through the text; in other words, we see an increase in explicit reference to how the text is to be used.

### 5.3 *Rhetorical Features: Direct Address*

Direct address first appears in the text at the same juncture as the first appearance of introductions and conclusions. Explicit reference to the user of the text, a putative astronomer, or *observer of the sky*, appears in different forms—nominal and pronominal—across subsequent component sections. The form of address stabilizes into the second-person pronoun in the later sections of MUL.APIN, a development that coincides with an increase in text coherence and complexity. A prominent related feature is the number of verbs predicated of the astronomer, which increases as the related astronomical material becomes more complex.

*Early*
There is no direct address in the earliest four component sections of the treatise.

*Middle*
The *ziqpu* star texts (subsections e-1, e-2) include the first references to the user of the text, in both in pronominal and nominal form (4.4.1.4.2, 4.4.2.4.1). As noted above, these are the first component sections that manifest some form of introduction and/or conclusion. By addressing the user of the text directly, the compilers of MUL. APIN identify the reader to whom the text is directed. Importantly, they also locate a deictic center, or, the perspective from which the astronomical phenomena under discussion are expected to be viewed. The canonical front of the *observer of the sky* is stated to be the point in relation to which the frame of reference is understood:

I iv 2 *opposite the breast of the observer of the sky*
I iv 8 *opposite your breast*

In section e, the pronouns "he" and "you" appear with a single verb predicate, "observe."

I iv 3 *he observes*
I iv 8 *you observe*

In section g, the first intercalation scheme, the pronoun "you" in the summary statement appears with two verb predicates: "observe" and "find" (calculate) (II i 23–24). The increase from one to two predicates is consistent with the overall increase in text complexity from sections e to g.

*Late*

The late sections of the treatise show stabilizing and conventionalizing of direct address. In sections j-2 and L, the putative astronomer is referred to only by use of the second person pronoun. This is consistent with the wider text conventions of the technical handbook tradition,[1] and represents a narrower range of reference types than that found in the middle portions of the text (noun; and third- and second-person pronouns).

In addition to the conventionalizing of form of address, the second intercalation scheme (subsection j-2) pairs the second person pronoun with seven verb predicates: "look, name, compute, predict, proclaim, find," and "multiply" (4.8.1.4.2). The number of verbs predicated of the astronomer, or user of the text, is consistent with an increase in both the complexity and precision of section j as a whole. Subsection j-2 consists of a series of generalized or axiomatic expressions that state rules for assigning years to types, "leap," or "normal." Each entry explicitly expresses the goal of the calculation.

In summary, the rhetorical features of introductions and conclusions, explicit mention of goals and procedures, and direct address of the user of the text, cluster together. The early sections have virtually no rhetorical features, with the exception of the "titles" or summary category statements; the middle sections show both introductory and concluding statements that mention goals and procedures, and a range of terms referring to the user of the text; the late sections are most coherent, employing conventionalized second-person form of direct address (along with numerous verbs specifying actions of the astronomer), and concise statements of goals and procedures.

## 5.4 *Natural Categories: An Emerging Taxonomy of Stars*

Stars and planets are referred to throughout the treatise. The primary content of the MUL.APIN treatise is the categorization of these "natural kinds," the description of their properties, and the definition

[1] We discuss this issue in Chapter 8, 8.1.

of parameters governing their appearance. The way in which they are categorized, however, can be seen to change through the different component sections of MUL.APIN.

We noted above (at the beginning of Chapter 4) that the Akkadian sign DIŠ seems to mark units of sense in the text, and we refer to it in the following summary of categorization phenomena. In many of the cases discussed below, the units of sense are also separate line entries (see below, 5.7, for additional discussion of DIŠ).

*Early*

In the early portions of the treatise, DIŠ marks individual star names (individuals or constellations) which usually appear as separate entries. Exceptions are physical lines which continue discussion of a star(s) from the preceding line, for example:[2]

I i 16 ¶ *The star which stands in the cart-pole of the Wagon:*
I i 17 *The Fox, Erra, the strong one among the gods.*

In cognitive terms, the star lists in section a describe basic object categories. Linguistically, the entries name individuals. The individual entries of the lists also constitute the members of the category named in the summary statements of each list. In logical, scientific, or mathematical terms, the three lists form the sets of interest over which subsequent generalizations are made.

In the original Akkadian text, line entries that comprise the summary statements at the end of each list are *not* marked by DIŠ. This suggests a conceptual distinction made by the compilers of MUL.APIN between the individual entries that name basic objects in the body of the list (marked by DIŠ) and the name of the category to which the objects belong (unmarked by DIŠ) that are found in the summary line entries.

*Middle*

In the *ziqpu* star texts, subsection e-1, DIŠ marks a category of objects. That is, a group of star names, written over three physical lines on the tablet, is marked by DIŠ as a single unit of sense[3] (4.4.1.4.3). In this

---

[2] An anomalous set of lines is I i 34–38.

[3] 'Sense' here is used following Frege's (1892) distinction between the 'sense' and 'reference' of ordinary nouns, or names for things, and does not refer to complete logical propositions underlying sentences—clearly, the entry marked by DIŠ here does not form a complete sentence.

intermediate component of the treatise, then, an entire group of stars, rather than separate individual entities within a list of basic objects, is marked by DIŠ (¶) as a single conceptual unit:

I iv 4 ¶ ŠU.PA, the star of Dignity, the Standing Gods, the Dog,
I iv 5 the She-Goat, the Panther, the Stag, the Old Man, the Crook,
I iv 6 the Great Twins, the Crab, the Lion, Eru, and the Abundant One.

In the same subsection e-1, a change can be observed in the wording of the summary statement, which now gives a full and explicit statement of the category name. In formal terms, it is an *ostensive* definition of the category, or, one which *points out* the category using the demonstrative pronoun "these," which refers back to the items in the foregoing list, rather than giving a linguistic definition:

I iv 7 *All these are the ziqpu stars*

The summary entry I iv 7 is not marked by DIŠ, like the earlier summary statements to the lists in section a, and again suggests that the compilers of the treatise made a conceptual distinction between the name of the category and the individual stars within it.

*Late*

In subsection j-2, a hierarchical, taxonomic category, or a "group of groups" (4.8.1.5) is found in a single text entry. The three categories of stars named earlier in the treatise, in section a, and there listed individually in the star lists Ea, Anu, and Enlil,[4] appear as a single sense unit:

II ii 7 *You lo[ok(?)] for the risings (?) and... of the stars of Ea, Anu and Enlil*

This entry occurs in the summary statement of section j-2, as part of a mathematical scheme that immediately precedes a mathematical procedure (4.8.1.4.4). We thus observe that the first articulation of hierarchical category structure co-occurs with the most mathematically-formal discourse up to that point in the treatise.

The summary is itself divided into two parts by the horizontal ruling following MUL.APIN II ii 10. The second part, MUL.APIN II ii 11–12, opens with DIŠ identifying these two lines, which conclude

[4] Similar statements are found in Enuma Anu Enlil, but in a different context; we discuss the distinction in Chapter 6, 6.3.5; and Chapter 7, 7.7.

that each lunar year is 10 days shorter than the stellar year, as a conceptual unit. One would expect DIŠ at MUL.APIN II ii 7 as well, although the Hunger-Pingree edition does not indicate that this is so (Hunger and Pingree 1989:93). Yet, this is not conclusive, as in all four extant manuscripts, the start of this line is either completely lacking or broken.

We should also note that the general category term "stars" is used earlier in the treatise, in the intermediate section (4.3):

I iii 49 *The stars enter into the night in the morning 1 UŠ each day*

The two lines in this section represent the most coherent and general individual entries found in the early portion of the treatise. Still, "stars" in these entries, while a general and presumably inclusive category name, is not marked in the text as a "group of groups," as we see in II ii 7. Thus, there is no way of determining the extension of the term, that is, the particular class of entities to which it refers.

It may be significant that the role of naming in category formation is explicitly noted in the text of the MUL.APIN treatise, in the lines immediately following the articulation of the general category year into the subordinate categories leap year vs. normal year:

II ii 8 *name this year*
II ii 10 *proclaim this year*

As discussed in Chapter 4, 4.8.1.5, this process of naming is expressed once as "naming" (the Akkadian verb *nabû*, II ii 8) and once as "proclaiming" (the Akkadian verb *qabû*, II ii 10). Naming plays a pivotal role in category formation, both at the basic level of category formation and the later emergence of constructed, or abstract, categories.

## 5.5 *Procedures and Procedural Categories*

MUL.APIN has many characteristics of the technical handbook tradition and describes numerous kinds of procedures. These procedures fall into types. References to procedures indicate to the reader how the information in the text is to be used, or interpreted. In other words, the compilers of the treatise increasingly set the astronomical information into an interpretive context that can be defined by the ultimate objective of the text: the practice of astronomy.

*Early*

The early component sections of MUL.APIN contain no procedural information.

*Middle*

The *ziqpu* star texts in section e (4.4) contain the first explicit references to procedures.

| | |
|---|---|
| I iv 8 | *...by means of which you observe* |
| I iv 9 | *the risings and settings of the stars at night* |
| I iv 10 | *If you are to observe the zipqu, you stand* |
| I iv 11 | *in the morning before sunrise* |

Each of the component subsections, e-1 and e-2, outline a *single procedure*. The procedure is described in an introduction and/or conclusion, and the stars being observed are given in the central portion of the component section. This single procedure is arguably the most basic performed in astronomy, or indeed, any other science: observation.

The first intercalation text (4.6) also contains generalized, procedural information in the conclusion, or summary portion of the text:

| | |
|---|---|
| II i 23 | *you observe the risings of the Sun, the visibility time of the Moon, the appearances of the Arrow,* |
| II i 24 | *and you will find how many days are in excess.* |

This entry designates the procedures common to the foregoing entries in the text, observation, and "finding," that is to say, calculating on the basis of the observations.

*Late*

Subsection j-2, the second intercalation text (4.8.1.1) concludes with a specific, general statement of the goal of the observations and calculations. Each lined entry within subsection j-2 consists of an "if-then" statement[5] that specifies both the conditions and goal of the procedure (4.8.1.3):

| | |
|---|---|
| II ii 1 | *[If] the Stars and the Moon are in conjunction [on the 15th of Arahsamnu], this year is normal.* |

[5] See Rochberg's (2009, in press) analysis of conditionals in the cuneiform corpus, and "if-then" structures in the protases and apodoses of the omen series in particular, and their overall significance in the emergence of science.

Each lined section contains the observed phenomena (in this case, stars and moon) the conditions (in conjunction) and the goal of the procedure (determining whether the year is a leap year or a normal year).

The final astronomical-mathematical section of the treatise, section L, is a comprehensive series of measurements for the length of day and night for each month. In section L, the goal of the procedure is stated in the conclusion rather than in each ruled entry as in subsection j-2. However, it differs from earlier component sections in the level of specificity, and in the highly abstract nature of the concluding portion of text which contains a formally-expressed mathematical axiom.

In summary, the general progression in procedural expressions through the component sections is thus from none, to single, to multiple and/or complex specification of procedures. In the late sections, where multiple entries relating to procedures occur within a single component section, they are generally of the same type and are expressed in parallel syntactic forms. This format appears to indicate emergent procedural types, or categories. As noted throughout the text analysis in Chapter 4, procedures do not yield to taxonomic classification as easily as objects do, and hence, subsection j-2 contains no summary statements that "name" the procedural type, or category.

## 5.6 *Definitions and Stipulation: Non-Natural Categories*

MUL.APIN contains a number of definitions and statements of equivalence, although none of them are purely stipulative, in Davidson's (1990) sense. That is, they all bear verifiable relations to objects and events in the world. Most are concerned with measurement: specifying units of time and distance as measured by water clocks and gnomons. Some definitions are procedural. For example, an "intercalary month" or a "leap year" are differentiated from ordinary months and years by stipulating their relevant defining properties. In MUL.APIN, these stipulated definitions are largely procedural. In the MUL.APIN treatise, there is a general progression from simple observed kinds (e.g. "stars") toward complex stipulated kinds (e.g. "intercalary month").

*Early*

In section a, the summary statements, at the conclusions of the star lists, are in the form of implied ostensive definitions (see 3.5). In other words, this entry is a statement of equivalence in which the definition is "pointed out" rather than linguistically expressed, using the demonstrative pronoun "these." For example, the summary statement from the end of the list of Enlil stars:

I i 39 *33 stars of Enlil*

has the implied meaning:

(These are the) *33 stars of Enlil.*

A fully-expressed ostensive definition is found later in the text, at the end of *ziqpu* star list:

I iv 7 *All these are the ziqpu stars*

A full statement of equivalence, where two linguistic expressions are linked by the use of the copula "is" (or implied use, as in the case of Akkadian) can be found in section b (4.2.4.1.4) in a single reference to *minas* as measured by the water clock:

I ii 43 *4 minas is a daytime watch, two minas is a nighttime watch.*

Reference to minas and to other non-natural, or stipulated measurement categories, such as UŠ, or degree, appear infrequently in the early portions of the treatise, and without being formally defined (4.3.4):

I iii 49 *The stars enter into the night in the morning 1 UŠ each day.*

*Middle*

The middle component texts are marked by an increase in the frequency of use of measurement terms, which are constructed, or stipulated, categories. In the first intercalation scheme, section g (4.6), both *mina* and NINDA, a measure of rate, are used in a single lined text segment:

II i 10 *4 minas is a daytime watch, 2 minas is a nighttime watch.*
II i 12 *(the sun) keeps moving... at a rate of 40 NINDA...*

*Late*

The final portions of the treatise are marked by a continuing increase in the frequency with which stipulated categories are used, and also of more complete definitional forms. The second intercalation text consists of a series of procedural definitions for categorizing years as normal or leap (as discussed in 4.8.1.5):

II ii 2 *[If] the Stars and the Moon are in conjunction on the 15th of [Kislimu], this year is a leap year.*

Section k is comprised of a table of statements of equivalences, in which specific values are given for terms of measurement (4.9). The statements of equivalence are achieved by the parallel placement of expressions on either side of the line of text (see 4.2.4.1.4 on the implied copula "is" in Akkadian, which may be relevant here).

| | | |
|---|---|---|
| II ii 22 | *1 cubit of shadow* | *2 ½ bēru of daytime* |
| II ii 23 | *2 cubits of shadow* | *1 bēru 7 UŠ 30 NINDA daytime* |

A formal definition occurs in section L, the final mathematical component of the treatise (see 4.10.5):

II iii 13 *4 is the coefficient for the visibility of the Moon*

As noted, none of the definitions in the text are purely stipulative, in Davidson's (1990) sense (see 3.5.1). That is, none of the terms defined are purely theoretical, fixed entirely by linguistic definition, of the type used in philosophy or theoretical mathematics. All bear some relation to observed states of affairs in the world. However, the definition in II iii 13, above, comes closest to the form of a purely stipulative definition.

### 5.7 *Ancient Forms of Text Marking: DIŠ and Horizontal Rulings*

Throughout MUL.APIN the editors of the canonical series utilize the stroke DIŠ and horizontal rulings as a sort of internal punctuation system. We noted above[6] that DIŠ is used to mark entries in the text. These entries often, but not always, consist of a single physical line of text. Horizontal rulings are most often used to separate sections and subsections of text. DIŠ and the horizontal rulings are not always used in what would appear to the modern reader to be a consistent and understandable pattern, but an overview of the treatise allows the following observations:

[6] See 5.4, above; Chapter 4, introductory paragraphs, and Chapter 6, 6.3.2, for additional discussion of DIŠ.

*Early*

At the beginning of the MUL.APIN treatise, entries marked by DIŠ correspond fairly consistently to the physical lines of text on the tablet. The notable exception is the summary statements that name the categories of stars (I i 39; I ii 18; I iii 35), which quite consistently are not marked by DIŠ. The first generalizations in the treatise (I iii 49–50) are also not marked by DIŠ.

*Middle*

As with other features of the text, the pattern evident in the use of DIŠ changes at section e, the *ziqpu* star texts. First, subsections e-1 and e-2 are marked by continuous, flowing discourse, rather than lists. Second, the units of text between the line rulings are larger, consisting of several physical lines of text.

DIŠ here appears at the beginning of the units demarcated by the horizontal rulings, and not necessarily at the beginning of each physical line of text. That is, DIŠ now corresponds with larger conceptual units of exposition rather than the physical line of text. Again, in subsection e-1, the summary statement is unmarked by DIŠ. The size of text units demarcated by horizontal units and by DIŠ shows more variation in the middle portion of the treatise. Two component sections are unmarked in their entirety (f-1 and h-i-2; found at MUL.APIN lines I iv 31–34, and II i 38–43); and several very long segments are marked by a single horizontal ruling and single DIŠ marker (II i 25–31; II i 38–43).

*Late*

From subsection j-2 on, the use of DIŠ and the horizontal line rulings seems somewhat more stable. The size of the text units marked by DIŠ seems more consistent, and they appear to possess greater conceptual coherence.[7] DIŠ is used to mark entries consisting of either single or multiple lines (II i 68–71), while summary statements, as in earlier component sections, are usually unmarked (II ii 7–10). Entries that are unmarked by DIŠ may function as commentary on the entries marked by DIŠ.

[7] The use of DIŠ in section m is not relevant to this discussion; see Chapter 4, fn. 18.

We should note that our observations regarding DIŠ are very preliminary. There are anomalies in the use of DIŠ[8] which we cannot explain on the basis of our current analysis alone. It does seem that summary statements unmarked by DIŠ have a different status than entries marked with DIŠ, which would suggest that the difference is intentional on the part of the compilers of the text. Perhaps the unmarked entries are later additions to pre-existing source material. If so, these anomalies may reflect decisions taken by the authors, editors, or scribes in the redaction and copying of MUL.APIN.

However, in the absence of comparison exemplars of the MUL.APIN series that predate the canonical version, these ideas remain speculative. A conclusive discussion of the role of DIŠ, including its relation to horizontal rulings, awaits a more comprehensive, detailed, comparative study of the use of these text marking devices throughout the cuneiform corpus.

### 5.8 *Generalizations, Axioms, and Assumptions*

Generalizations occur throughout the treatise. Indeed, there would be little point in compiling the treatise if the content could not be generalized to recurring observation and measurement practices. The development through the treatise shows increasing completeness and explicitness of generalized expressions, and broader scope of the phenomena over which the generalization holds.

The first syntactically complete generalizations occur in the intermediate text following section d. Like many of the summary statements, they are unmarked by DIŠ:

I iii 49 *The stars enter into the night in the morning 1 UŠ each day.*
I iii 50 *The stars come out into the day in the evening 1 UŠ each day.*

The two statements found in this intermediate section are generalizations over the category "stars," which in these statements has an indeterminate extension. That is, the individuals referred to by the term "stars" are not specified, but are assumed to be known by the reader of the text.

[8] See, e.g. Chapter 4, 8.1.3, MUL.APIN II ii 7–17, where of three summary statements, two are unmarked, which appears to be the convention, while one is marked. There is no immediately evident reason why this should be so.

We refer to these generalizations as "proto-axioms" (see discussion at 4.3.3.2; 4.7.5.4.3), but don't belabor the point, as the distinction between generalizations and axioms, within the scope of our analysis, is arguable (Chapter 3, 3.6). In logic, an axiom is usually held to be a postulate, self-evident and not needing proof. In mathematics, an axiom is a statement that functions as a starting point from which other expressions can be logically derived. Axioms cannot be derived and cannot be demonstrated by mathematical proof.

In domains other than mathematics and philosophy, however, *axiom* can be used, in a non-logical sense, to mean *an assumption or any established principle generally held to be true within that domain*. In the broadest sense, then, the rhetorical and conceptual status of both generalizations and axioms can be understood only in relation to an accepted body of knowledge. In the case of MUL.APIN, they can be understood in relation to the background assumptions shared by the compilers and, in most cases, the readers or users of the text. Determining post-hoc what was held to be generally true for Mesopotamian astronomer-scribes is difficult, hence the difficulty in distinguishing a generalization from an axiom on the basis of textual evidence alone.

Using textual evidence as a criterion, the clearest example of an axiom in the treatise is the formal stipulative definition that is found at the conclusion of section L:

> II iii 13 *4 is the coefficient for the visibility of the Moon*

As noted above (see 5.6, above; and 3.5.1) a stipulative definition has the force of an assumption. In this sense, then, while allowing for the difficulty of ascertaining the underlying knowledge of the Mesopotamian scribes who compiled and used the text, II iii 13 is an axiom.

## CHAPTER SIX

# DISCUSSION: MUL.APIN, WRITING, AND SCIENCE

The MUL.APIN treatise was composed near the dawn of scientific endeavor. We began our analysis with the assumption that its composite form reflects processes underlying the articulation of a written science. That is, we assume that the treatise has an historical-cumulative aspect. We attempted to make our analysis as theory-neutral as possible by developing and applying a naturalistic analytic method, in an effort to open the text up to further analysis rather than restrict it to one interpretive perspective alone.

One of our goals, however, was to explore the text for evidence that writing may have played a causal role in conceptual change and the emergence of science. The uniqueness of MUL.APIN invites this question. If a window exists onto the role writing may have played in the early emergence of scientific thought, surely it is to be found in just such a text. A textual analysis alone cannot *prove* that the medium of writing *caused* conceptual change in the astronomer-scribes who employed it. Claims to this effect would also seem tenuous in light of our cultural-historical remove from the Mesopotamian era in which the text was composed.[1] We therefore embark on the following discussion in the spirit of reasoned speculation rather than objective proof.

### 6.1 *A Developmental Progression*

The pattern of observations summarized in Chapter 5 is consistent with a developmental progression. Both the conceptual content of the MUL.APIN treatise, and the form of written expressions by which that content is expressed, become increasingly complex and sophisticated as it progresses from early to late component sections. This pattern is common in the cuneiform corpus. The scribal curriculum itself

[1] Still, the historical gap is by no means insurmountable; see discussion in Chapters 1 and 2.

shows a simple to complex progression (Veldhuis, 2004), and other, earlier texts show a similar pattern, from simple to complex. The sort of detailed analysis we present here is not necessary to make this point, since one need only visually examine the text to see that later sections are more complex than the early ones.

Such a visual examination might suggest a straightforward explanation of the developmental progression in MUL.APIN: it reflects the additive nature of the cuneiform text tradition, in which content is preserved and added to rather than supplanted or replaced by newer material. We might, then, appeal to something like an accretion model to account for the developmental progression. On an accretion model, MUL.APIN would illustrate the development and accumulation of skills and knowledge. In other words, the astronomer-scribes who compiled the later component sections were better writers, and more knowledgeable astronomers, than those who compiled the earlier sections. In fact, they were probably both. One could hardly expect that in the process of accumulating and recording their observations, they would become *less* skilled at exposition or *less* knowledgeable about their field.

However, there are at least two problems with an accretion model. First, the later component sections do not incorporate all earlier developments, but rather, are different in form and content. They are more conventional, explicit, and procedurally complete, and the complex content they contain is expressed in more coherent language. The changes through the treatise are thus qualitative rather than simply additive.

Second, the changes that appear through the treatise are neither steady nor monotonic, but rather occur in "bursts." These bursts of change result in a "wobble" in the developmental progression through the successive component sections of MUL.APIN, which suggests that the astronomer-scribes were not simply *accumulating* knowledge and skills, systematically adding new to old, but were instead engaged in a more dynamic process. An accretion model cannot explain this pattern, but it is consistent with the inferential model of the type we offer in Chapter 2, 2.4.

## 6.2 *Applying an Inferential Model to MUL.APIN*

A progression toward a more conventional, explicit, and procedurally complete text suggests an increasing grasp of requirements on text interpretation. Any text necessarily creates the possibility of isolated readers. Texts are physical objects that can be transported and read in contexts not anticipated by their writers. Reading a text under these conditions could have led to misinterpretations that a more explicit text would have prevented. Improving a text to take this into account would indicate that the composers had experienced and reflected upon the difficulties inherent in text interpretation, and altered the form of written expression accordingly.

Isolated readers, of course, were not the norm in the Mesopotamian scribal tradition. Canonical MUL.APIN was composed and interpreted within a community of scribes who shared oral and written traditions.[2] A typical example would be the royal astronomers of the 7th-century Assyrian court, who reported their findings in written form, often citing interpretations of omens from the Enuma Anu Enlil series that were based on oral authorities (Hunger 1992). Shared knowledge within the oral tradition was thus a ready source of information available to any scribe, or "reader" of the text, when difficulties in interpretation arose. Still, occasions must have arisen in which an astronomer-scribe found himself alone with a tablet, a text that had to be interpreted in the absence of his peers.

Explicit texts have fewer ambiguities and lead to fewer misinterpretations. An isolated reader is reliant on what is on the page (or clay tablet). The richer contexts of spoken communication, in which non-linguistic sources of information abound, are absent. When the reader doesn't understand what is written, the only recourse is to infer what the writer may have meant from what is written. Linguistic form then guides interpretation and re-interpretation, and the inferential environment is thereby *recalibrated*, or biased toward what can be represented in writing: the more precise the information in the text, the more accurate its interpretation. Consistent and accurate readings are thus enhanced by explicit texts.

Direct evidence of this in the text would manifest in altered uses of linguistic expressions that are dependent on non-linguistic context for their interpretation. Indexicals, for example, are terms that require context to be understood.[3] Spatial expressions such as *here* and *there, up*

---

[2] See the discussion in Chapter 2, 2.1; see also Stock, 1983; Olson, 1994, for manuscript literacy and oral traditions.

[3] See Chapter 3, 3.2, for a discussion of indexicals and their relation to context.

and *down*, temporal expressions like *now, then*, and *today*, motion verbs such as *come* and *go*, along with personal pronouns, such as *I, you, he*, are understood in relation to context: who is speaking, to whom, when, and where. They are substantially more difficult to interpret in written text unless the frame of reference for their interpretation is made explicit.

A transition from vague to specific indexical expressions would suggest that the astronomer-scribes were taking this into account. If indexicals in the text are increasingly comprehensible, precise, and interpretable, it would suggest that the scribes were increasingly aware of requirements on interpretation, or, how to make the text comprehensible to their intended readers. In other words, they were increasingly *writing in order to be read.* Evidence for recalibration may thus be found in altered use of indexicals, as would any feature of the text that clearly takes the reader into account.

A second aspect of the treatise in which this process is in evidence is rhetorical features. Introductions and conclusions indicate to a reader the purpose of a text. Similarly, direct address of the reader of the text would indicate that the astronomer-scribe was taking a reader into account. Second person address was a convention of the technical handbook tradition[4] in the cuneiform corpus, and it could be argued that its use simply reflects a scribal convention. But in MUL. APIN, the emergence of rhetorical features coincides with changes in the use of indexical expressions. This pattern of features suggests a causal account, that is, a reason for the emergence of the convention in the first place.

### 6.2.1 *Textual Evidence for Recalibration: Rhetorical-Indexical Clusters*

Rhetorical features pattern together with indexical terms in the MUL. APIN treatise. A comparison of the early, middle and late portions of the treatise illustrate how these two features develop through the component sections.

[4] This tradition is discussed in Chapter 8, 8.1.

*Early*

In the early component sections of the treatise, the star lists of section a, locations of stars and constellations are expressed using relative or intrinsic frames of reference. The indexical expressions used are minimally informative. The text does not specify the perspective from which the expressions are to be understood:

| | |
|---|---|
| I i 12 | *ŠU.PA, Enlil who decrees the fate of the land.* |
| I i 13 | *The star which stands in front of it: the Abundant One, the messenger of Ninlil.* |
| I i 14 | *The star which stands behind it: the Star of Dignity, the messenger of Tišpak.* |

The indexical terms "behind" and "in front of" are relative, as stars have no canonical front or back orientation. The indexical terms can only be understood as relative to the position of the viewer, or perhaps in rare cases with reference to the shape and/or orientation of the constellation itself. Yet no articulation of a deictic center[5] or perspective is given in the text. Spatial-temporal information in the early component sections is thus "under-marked." That is, the text supplies less information than needed for clear interpretation.

Similarly, rhetorical features are lacking. No introduction is given in the text that might serve to indicate how the star lists are to be read, or what purpose the observation of the stars is intended to serve. The text simply presents a list of stars with some identifying features and a summary statement that names the category of stars.

*Middle*

The first appearance of rhetorical features co-occurs with an altered pattern in the use of indexical terms. The *ziqpu* star text in section e, toward the middle of the treatise, manifest introductions, conclusions, and direct address of the reader of the text, along with new indexical expressions.

The spatial-temporal indexical terms used in e-1 are more comprehensive and precise than those in the early sections of the treatise, and use an absolute frame of reference: the cardinal directions, South, East, and West.[6] In contrast to the under-marking of section a, here,

[5] See Chapter 3, 3.2.

[6] These terms indicated a range rather than points on a compass to Mesopotamian readers (see Chapter 3, 3.1, for Levinson's notion of frames of reference); nevertheless they are fixed and objective.

*more information is given than is actually needed* to understand the perspective from which the stars are to be viewed:

I iv 10–13 *West to your right*
*East to your left*
*Your face directed towards South*
*In the middle of the sky*
*Opposite your breast*

If an observer is facing South, East and West are always in the same relative orientation and don't need to be specified in the text. There is thus a redundancy, or "over-marking," of spatial-temporal indexical terms in subsection e-1.

Rhetorical features also appear in subsection e-1. *Introductory* and *summary* statements specify the goal of the text, that is, the observation of the *ziqpu* stars. The text also states the position from which the observation is to be carried out. The reader of the text, or "the observer of the sky," is also directly addressed. This identification of the observer serves to identify a deictic center, or perspective, from which the spatial-temporal indexical terms are to be understood. The reader of the text is identified using multiple referential forms: a nominal phrase, "the observer of the sky," and both second- and third-person pronouns "you" and "he."

Spatial-temporal reference includes absolute terms and is "over-marked": the text supplies more information than is needed for the perspective of observations to be clearly understood. There is thus an alteration in both *type* and *quantity* of indexical terms, from *less* than what is required for interpretation (in section a) to *more* than what is required (in subsection e-1). These variations are of the sort that Grice (1975) referred to as violations of the maxim of quantity: too much or too little information is being supplied.

Late

The variation observed in the later portions of the text is toward an *appropriate* quantity of information and *conventionalized* forms of expression. Spatial-temporal indexical terms increasingly rely on objective, or absolute, frame of reference in the later portions of the text. Cardinal directions appear most frequently, although they do not entirely supplant relative frames of reference. Multiple-marking of location and time still occurs, but the type of over-marking, or redundant marking, seen in the *ziqpu* star texts in subsection e-1 is not in evidence:

II ii 3 [¶ If ] *the Arrow [becomes visible] in the East in the evening on the 15th of Ṭebetu, [when (?) the Sun] rises towards the South, turns and keeps coming up towards the North, this year is normal.*

Rhetorical features increasingly take the form of explicit instructions that usually, but not always, occur in distinctly marked concluding sections, such as the one found in section j:

II ii 13 *If you are to find the correction for day, month, and year:*
II ii 14 *you multiply 1,40, the correction for a day, by one month,*
II ii 15 *and you find 50, the correction for one month; you multiply 50, the correction for one month,*
II ii 16 *by 12 months, and you find 10 additional days, the amount for one year.*
II ii 17 *In three years (variant: the third year) you proclaim (this year) a leap year.*

In the later portions of the treatise, direct address of the user of the text is conventionalized, and consistently occurs in one referential form only, the pronominal "you." This convention is consistent with that of the technical handbook tradition,[7] and contrasts with the multiple forms of address found in subsection e-1 in the middle portion of the treatise.

*The more complex content of the later portions of text thus co-occurs with increasing conventionality of expression and textual form.* More concise spatial-temporal indexical expressions conform more closely to the Gricean maxim of quantity (Grice, 1975; 1989). That is, they include the appropriate amount of information necessary to interpret the text.

Explicit statements of goals and procedures are expressed more consistently, in detailed concluding sections or in brief references within the body of the component section. The late sections of text are thus more easily interpretable, which renders the text more authoritative and autonomous. In other words, the form of the later texts suggests a greater independence from the oral tradition, since there is less need for consultation and prior knowledge in order to interpret it. These changes may reflect an increased awareness, on the part of the scribes, of the advantages conferred by completeness and explicitness in a text.

---

[7] See 1.4 and 2.1 for the cuneiform scribal tradition; and 8.1 for the technical handbook tradition.

### 6.2.2 *Summary: Rhetorical-Indexical Clusters*

The developmental progression through the treatise, with respect to indexical terms and rhetorical features, illustrates an increasing conformity with requirements on text interpretation. In the early components of the treatise, there is no indication of the frame of reference within which spatial indexical expressions are to be interpreted and no rhetorical features that indicate to the reader how to use the text. The early sections seem intended for use primarily within contexts of an oral tradition, because the text itself is *under-marked*.

In the middle portion of the treatise, a flurry of spatial-temporal indexical terms appears. More information is given than is required for interpretation. The text is redundant, or *over-marked*. This over-marking co-occurs with multiple forms of address. The reader is addressed in second- and third-person pronouns and a nominal form: "the observer of the sky." These multiple forms of direct address, along with the redundant use of indexical expressions, appear at the same time as new rhetorical features, introductions and conclusions, in the middle component text.

This overuse is not in evidence in the later component texts, where more conventional and systematic use of indexical expressions and form of address are the norm, and appropriately marked. The form of address is exclusively the second-person pronoun. Spatial indexicals are used almost always in relation to absolute frames of reference. Goals and procedures are consistently and repeatedly expressed. Instructions to the reader are more complete and explicit.

The final component texts, then, are much closer to striking the right balance for text interpretation. Increasingly complex content is expressed in an increasingly conventionalized and concise form. Concluding sections, which detail procedures, calculations, and the intended goal of the procedure, appear more consistently.

From beginning to end of the treatise, then, a "wobble" in the developmental progression can be observed. It begins with under-marked and incomplete text, then progresses through over-marked or redundant expressions, and eventually ends with appropriately-marked, consistent, and conventionalized expressions. An accretion model that appeals to simple accumulation of knowledge or writing skills can't account for these qualitative changes, nor can it explain the "wobble" in the progression. The patterning of rhetorical and indexical features indicates a more dynamic, epistemic process, a process of trial and error toward more sophisticated understanding of what makes a text more comprehensible to a reader.

The changes in the use of indexical terms through the treatise—from under-marking to over-marking to appropriate marking—are consistent with acquiring the Gricean *maxim of quantity*. That is, the text should include the right amount of information, not too little and not too much.

It is highly unlikely to be a coincidence that the over-marking of indexical expressions occurs at the exact juncture in the treatise where introductions and conclusions first appear. Both of these developments would indicate that the compilers of the treatise are taking interpretation into account. Early component sections—the star lists, for example—suggest interpretation within an oral scribal tradition. Indexical expressions are easily understood when it is clear who is speaking, when, and where, and extra-linguistic sources of inference abound. In the later component texts, in contrast, the assumptions, or frame of reference is more explicit, making the text more easily interpreted on its own.

The later sections of MUL.APIN thus represent not only an accumulation of knowledge, but also suggest that the astronomer-scribes who composed them were more aware of the requirements of text understanding. Indexical terms become more interpretable, and at the same time, rhetorical features appear, and complex content is expressed in an increasingly complete and conventional manner. All of these developments allow more reliable interpretation. The appearance of rhetorical-indexical clusters in the text, then, is highly consistent with an inferential model in which writing *recalibrates* the interpretation process.

Is there any evidence that this process leads to broader conceptual change? We now consider whether the appearance of category organization, generalizations, and definitional form in the treatise changes concomitantly with the above pragmatic indicators of recalibration. Do the later component sections of MUL.APIN show an increase in logical, rational thought?

## 6.3 *Textual Indicators of Logic and Rational Thought in MUL.APIN*

Category organization is part of universal core cognition, but it is measurably and reliably influenced by both culture and expertise.[8] There is every reason to expect, then, that a cultural invention such as writing, and its regular and repeated use, would have an effect on category organization. Since identifying and articulating taxonomies is a central activity of the observational sciences, we looked for evidence in MUL. APIN of incipient hierarchical category organization.

### 6.3.1 *An Incipient Taxonomy of Stars*

The star lists in section a of the MUL.APIN treatise identify three distinct "categories" of stars in sequence. Semantic, or conceptual, coherence can be found in earlier exemplars of lists, such as the encyclopedic Urra = *hubullu* series, and is not unique to MUL.APIN. However, the placement of the lists in this composite treatise allows them to be considered together with subsequent component sections, broadening the context of their interpretation. In effect, as our analysis in Chapter 4 (4.1.4.1) suggests, the star lists identify the set of objects that form the subject matter of the treatise.

Early in the treatise, the *individual line entries* of the lists in section a identify individual stars, or *basic level objects.*

I i 16 ¶ *The star which stands in the cart-pole of the Wagon:*
I i 17 *The Fox, Erra, the strong one among the gods.*

The summary statements of each one of the three star lists *name the category* to which the individual stars in the foregoing list belong, that is:

I i 39 *33 stars of Enlil*
I ii 18 *23 stars of Anu*
I ii 35 *15 stars of Ea*

[8] See Medin, Lynch & Solomon, 2000; Medin & Atran, 2004; Atran, Medin, & Ross, 2005, on categories and culture; Bowerman & Levinson, 2001, Gentner & Goldin-Meadow, 2003, for cross-cultural and developmental research showing the influence of culture and language on cognition; and Tomasello, 1999, for a general account of culture and cognition.

In the middle portion of the treatise, in the *ziqpu* star text in section e-1, an individual line entry identifies all the stars in the category. Rather than a series of individual separate line entries, one for each separate star name, the *ziqpu* stars are thus grouped in a single entry. We see, then, a single continuous linguistic expression that articulates the entire category of stars:[9]

I iv 4 ¶ *ŠU.PA, the star of Dignity, the Standing Gods, the Dog,*
I iv 5 *the She-Goat, the Panther, the Stag, the Old Man, the Crook,*
I iv 6 *the Great Twins, the Crab, the Lion, Eru, and the Abundant One.*

The summary statement at the end of subsection e-1 *names the category* in a complete sentence:

I iv 7 *All these are the ziqpu stars*

In the late portions of the treatise, in subsection j-2, an individual line entry names the three categories of stars:

II ii 7 *You lo[ok(?)] for the risings (?) and... of the stars of Ea, Anu and Enlil*

The three categories of stars named here constitute a single, inclusive, taxonomic category, or, a "group of groups," that includes all the stars named separately in the three star lists of section a.[10]

There is thus a logical progression through the treatise toward increasingly comprehensive categories, from individual members, to groups, to a "group of groups." A parallel progression is also evident in the linguistic form of the expressions that *name* the categories of stars. In the star lists of section a, the three categories of stars are named separately at the end of each list in incomplete sentences. In the middle portion of the treatise, in subsection e-1, the name of the category (the *ziqpu* stars) is expressed in a complete sentence in a summary statement that forms a conclusion. In the late portion of the text, the individual line entry names three star groups in a single complex sentence that forms a concluding statement at the end of subsection j-2.

[9] The "punctuation" mark DIŠ (¶ in the text) identifies a single entry, see discussion below, 6.3.4.

[10] The significance here is in the placement of the phrase, after the lists and generalizations; that is, the statement here implies the prior content of the treatise. We discuss this in relation to the cognitive function of writing in MUL.APIN, Chapter 7, 7.6.

Thus, while the MUL.APIN treatise may not present a fully developed system of logical classification that, from a modern perspective, distinguishes and exhaustively categorizes all known heavenly bodies, it does reflect the articulation of an incipient taxonomy of stars.

### 6.3.2 *Generalizations*

The identification of individual stars, and the naming of categories to which they belong, is immediately followed by *generalizations over the categories*. Sections b–d give the temporal parameters governing the appearance of individuals, for example:

I ii 36 ¶ *On the 1st of Nisannu the Hired man becomes visible*

The temporal parameters on the visibility of individual stars and constellations are then followed by a *generalization over a class*, (*the stars*) given at I iii 49–50:

I iii 49 *The stars enter into the night in the morning 1 UŠ each day.*
I iii 50 *The stars come out into the day in the evening 1 UŠ each day.*

There is thus a logical progression in the text, from the particular to the general: first, individuals are named, then the category to which the individuals belong; and then *generalizations over the category* are given.

### 6.3.3 *Generalizations and the Text Marker DIŠ*

Most individual line entries throughout the treatise are marked by the "punctuation" mark, DIŠ (rendered ¶ in the transliterated text here). Single entries usually comprise a single physical line of text, but some entries span several physical lines. The absence of DIŠ from specific entries, then, may be meaningful.

In section a, for example, DIŠ marks entries that name individual stars, while the summary statements that name the category of stars are not. Other entries *not* marked by DIŠ include the the two generalizations at I iii 49–50, above, and most of the summary statements throughout the treatise. DIŠ seems to be absent from general statements that refer back to, or comment on, individual entries in the preceding text.

Its use, then, may indicate a reflective stance by the astronomer-scribes toward what they were writing about. They may have recognized different sorts of entries as distinct. There is, of course, also the possibility that the absence of DIŠ simply indicates later additions to the text. Further study across a wider sample of texts would be necessary before any conclusions about the functions of DIŠ can be drawn with any confidence, but on our view, its use in MUL.APIN suggests a distinction between *text* and *commentary*, or a conceptual distinction between *observation* and *generalization*.

### 6.3.4 *Definitions: Content and Form*

Definition is at the heart of the scientific enterprise, and describing properties, or, more accurately, listing criterial features of categories, is at the heart of the process of definition. It is conceivable, even plausible, that *writing a series of entries* that identify the individuals in a particular group, or category, would generate reflection on the characteristic features or properties that they all shared. That is, the astronomer-scribes may have asked themselves: what sort of criteria would lead these individuals to be included in this category?

The developmental progression through the MUL.APIN treatise shows two sorts of changes related to the emergence of definition. One change is in the type of *content*, or the nature of the categories being defined. The second sort of change is in the *linguistic form* that the definition takes.

#### *Content*

In the early part of the text, stars—an observed natural category of phenomena—are identified and named. Later in the treatise, *constructed categories*, or artifacts of human activity, are much more in evidence. The water clock and gnomon, units of measurement such as minas, UŠ, NINDA, *bēru*, along with stipulated definitions that pertain to them, appear with greater frequency in the later portions of the text. For example, the constructed category "leap year" appears in the second intercalation scheme:

> II ii 2 *[If] the Stars and the Moon are in conjunction on the 15th of [Kislimu], this year is a leap year.*

"Leap year" is not the same sort of category as "star." Forming the conceptual category "star" requires only direct apprehension, while "year" requires continuous observation over time and, to some extent, calculation. "Leap year" is a further articulation of the stipulated category "year," that is, certain conditions must hold in order for a year to fall into the category "leap year."

Measurements, procedures, and the like are more in evidence in the latter part of the treatise, as are numbers and calculations. Section k, for example, is composed almost entirely of measurements.[11]

*Linguistic Form: Ostensive Definition*

The simplest form of definition is *ostensive*, in which an indexical expression "that" or "this," or even a manual point, is used to denote an object or class of objects. The expressions "That is a zebra," for example, or "All these are apples, all those are pears," use the indexical expressions *that, these, those* to denote a kind or class of objects without a linguistic definition of the words *zebra, apple*, or *pear*.

Early in the MUL.APIN treatise, statements of category membership take the form of an *implied ostensive definition*.[12] The three star lists in section a conclude with statements that denote the categories to which the foregoing stars belong, although the indexical pronoun is implied rather than stated:

I i 39 (All these are) *33 stars of Enlil*

In the middle portion of the treatise, in section e-1, the summary category statement takes the form of an *explicit ostensive definition*:

I iv 7 *All these are the ziqpu stars*

[11] Numbers are generally held to represent a distinct domain within human cognition, with conceptual status different from that of observed natural categories, in which names for objects are bound to particular conceptual and ontological domains of reference. Numbers, in contrast, can be applied to any domain of objects, actual or hypothetical. In the cognitive development of children, the domain of number has a path of development that is distinct (Spelke & Kinzler, 2007; Spelke & Tsivkin, 2001). Exact numerical concepts for numbers larger than four and the construction of exact, complex systems of numerosity is generally held to involve an inductive leap of some kind, possibly involving language (Carey, 2001). The conceptual content of the latter part of the treatise, from this perspective alone, is thus more abstract than the earlier part.

[12] See discussion of forms of definition in Chapter 3, 3.5.

*Linguistic Form: Statements of Equivalence*

The most familiar linguistic form of a definition is a statement of equivalence,[13] where the copula "is" (or in the case of I iv 7, "are")[14] links a *definiendum*, or term to be defined, and a *definiens*, or defining linguistic expression (see 3.5). In dictionaries, the two terms usually appear in apposition, one simply following the other. In either case, some degree of semantic equivalence is implied.

In the MUL.APIN treatise, statements of equivalence emerge *early* in the treatise, in section b:

II ii 10 *4 minas is a daytime watch, 2 minas is a nighttime watch.*

The *frequency of statements of equivalence increases* through the treatise. Section k consists almost entirely of a set of expressions whose equivalence relations are achieved by parallel placement:

| | | |
|---|---|---|
| II ii 22 | *1 cubit of shadow* | *2 ½ bēru of daytime* |
| II ii 23 | *2 cubits of shadow* | *1 bēru 7 UŠ 30 NINDA daytime* |

*Linguistic Form: Stipulative Definition*

Stipulative definitions are used to determine the intension, or meaning, of a term for purposes of a particular theoretical argument or proof. For example, in geometry one finds definitions of the sort, a *circle is the locus of points equidistant from a given point.*

The definitions and statements of equivalence found in MUL.APIN all bear verifiable relations to observable phenomena in the world, and are thus not *purely* stipulative in Davidson's (1990) sense (see 3.5.1). "Month" and "year," for example, are determined by sustained observation over time, but are not defined in the treatise. The related notions of "intercalary month" and "leap year" are defined in MUL. APIN by the astronomical procedures in which they figure. This is illustrated by the concluding portion of component section j-2:

II ii 13 *If you are to find the correction for day, month, and year:*
II ii 14 *you multiply 1, 40, the correction for a day, by one month,*
II ii 15 *and you find 50, the correction for one month; you multiply 50, the correction for one month,*

[13] See 3.5.

[14] In Akkadian the copula is implicit rather than explicit; see 4.2.4.1.4.

II ii 16 *by 12 months, and you find 10 additional days, the amount for one year.*

II ii 17 *In three years (variant: the third year) you proclaim (this year) a leap year.*

The meaning of the term "leap year," while based on the simpler, or more basic, category "year," requires definition. However, the definition is not purely stipulative. That is, the meaning of the term is not entirely determined by language, since real-world concerns dictate the sorts of procedures involved in the assignment of a year to the category "leap." Nevertheless, definition—procedural or otherwise—is clearly at the root of articulating the meaning of the term.

The linguistic form in MUL.APIN that most closely approximates a purely stipulative definition appears in section L:

II iii 13 *4 is the coefficient for the visibility of the Moon*

Again, this is not a purely stipulative definition because there are real-world constraints on the visibility of the moon. Thus, the meaning of the expression is not determined by language alone. But the expression is a semantically complete logical form and, while not purely stipulative, it is formal, abstract, and axiomatic.

### 6.3.5 *Summary: Categories, Generalizations, and Definition*

There is a logical progression through the treatise from the particular to the general that manifests in a number of ways. Individual objects are first named, then identified as belonging to categories; in the middle portion of the treatise, the category of *ziqpu* stars appears in a single continuous line entry; and late in the treatise, a single line entry names all three of the earlier mentioned categories of stars (Ea, Anu, and Enlil stars), in a "group of groups," suggesting the formation of an incipient taxonomy.

Generalizations over the categories of stars appear only after the listing of category members and naming of the category. The treatise begins with simple ostensive definitions of observed natural kinds (stars), while later sections of text give complex procedural definitions of stipulated, or constructed, categories, such as measurements, "leap years," and "intercalary months." Toward the end of the treatise, an axiomatic expression appears in a formal statement of equivalence.

There are thus indications in the text that the conceptual content of the later portions of MUL.APIN is of a higher order, and more abstract, than that of the earlier part. This conceptual change co-occurs with changes in textual indicators of recalibration, primarily rhetorical-indexical clusters. However, further comment is necessary regarding two line entries that we have argued are significant here, the "group of groups" at II ii 7 *You lo[ok(?)] for the risings (?) and... of the stars of Ea, Anu and Enlil*, and the axiom at II iii 13 *4 is the coefficient for the visibility of the Moon.*

We have noted that similar statements appear elsewhere, and much earlier, in the cuneiform corpus. Stars of the paths of Anu, Enlil, and Ea are mentioned together in the Astrolabes and in the omens of Enuma Anu Enlil, and coefficients (*iggigubû*) such as the one for the visibility of the moon appear grouped together in list format in mathematical texts (Robson, 1998). Their significance in MUL.APIN, on our view, relates to their placement within the treatise. A clearer picture of how this might work requires some elaboration of the cognitive function of writing, with particular reference to current dual process models of cognition. We address this in the following chapter.

CHAPTER SEVEN

# FURTHER THOUGHTS: THE COGNITIVE FUNCTION OF WRITING IN MUL.APIN

There is no reason, in principle, why scientific theories should be off-limits to any human culture, even pre-literate ones. Developmental evidence now suggests that scientific theories develop out of the core conceptual knowledge common to every human being (Carey & Spelke, 1996). Members of traditional, pre-literate cultures have been shown to possess core knowledge even in abstract domains such as geometry,[1] and it is thus conceivable that advanced scientific theory building, including the stipulation of new categories and terms, *could* have developed without writing. The question is *would* it have?

The earliest uses of writing were independent of spoken language.[2] As Cooper (2004:83) points out, writing was initially intended for uses in areas where spoken language couldn't do the job. Otherwise there would have been no need to invent it. Once created, its uses multiplied and its forms developed. The cuneiform corpus provides a unique window onto these early developments, and MUL.APIN, to some extent, mirrors the progression of written forms in the cuneiform corpus. It begins with the earliest extant written form, the list. Subsequent component sections include complete sentences in connected discourse. Astronomical content and textual form appear to develop together in a reciprocal manner. In order to understand how this might happen, it is necessary to consider the broader functions of writing in cognition.

Olson (2001) has argued that the opacity of written forms inherently conveys a reflexive attitude toward the content that writing conveys. This property of written signs, to simultaneously represent *content* and *attitude*, is central to explicit thought (Dienes & Perner, 1999; Evans & Over, 1999). The developmental progression observed through the successive component sections of MUL.APIN is consistent with

[1] Evidence for this can be found in the empirical studies of Dehaene, et al., 2006; cf. also Spelke & Kinzler, 2007.

[2] See Schmandt-Bessarat, 1986; Nissen, Damerow & Englund, 1993; Olson, 1994, 1996.

this sort of effect. The increasing completeness and coherence of the text—or, increasingly explicit textual form—co-occurs with evidence that the compilers of the treatise were increasingly writing in order to be read. Rhetorical-indexical clusters (6.2.1) and the use of the text marker DIŠ (6.3.3) both suggest a developing distinction between *content* and a *reflexive attitude toward that content.*

Current models of cognition provide a new lens through which to view this distinction. We first outline the models and then consider how they help to illuminate the findings outlined in the previous two chapters.

## 7.1 *Writing and Dual-Process Models of Cognition*

Dual process models of cognition have recently gained wide acceptance in the cognitive sciences.

> There is now abundant evidence for the existence of two types of processing in human reasoning...System 1 is a collection of autonomous sub-systems, many of which are old in evolutionary terms and whose operations are fast, automatic, effortless, non-conscious, parallel, shaped by biology and personal experience, and independent of working memory and general intelligence. System 2 is more recent, and its processes are slow, controlled, effortful, conscious, serial, *shaped by culture and formal tuition*, demanding of *working memory*, and related to general intelligence...System 1 being highly contextualized, associative, heuristic...System 2 being decontextualized, rule-governed, analytic.
>
> Frankish, 2009:89 (italics, ours)

There remains substantial argument over whether these differences are best characterized as separate systems or different levels of processes,[3] but the distinction itself is generally accepted. The dual process distinction clarifies the relation between spoken language and writing. Spoken language is universal, and thus must originate in, or at the very least be massively underwritten by, System 1 cognition. In contrast, the properties of writing outlined in Chapter 2 identify it as a product of System 2. Writing does an imperfect job of representing spoken language, and necessarily so. The auditory-vocal and ancillary properties of spoken language are more wide-ranging, subtle, and complex

[3] See Evans and Frankish, 2009, for a collection of diverse views.

than orthographies can represent, and for all we know, some aspects of the System 1 capacity for language may not be even available to consciousness.

Written language cannot, then, be reasonably understood as "the same as" spoken language, a representation that has been simply transposed from a vocal-auditory modality into a visual-spatial one. Written language may be better understood as a *System 2 construal of language*. Such construal can take a wide range of forms: alphabetic, syllabic, logographic, hieroglyphic, and so forth. These forms have variable correspondence relations to spoken language. The adequacy criterion for orthographies is thus not the "verisimilitude" with which they represent language, which must always be imperfect, but rather the ease with which they enable the *recovery of logical form*. To be of use, writing must convey meaning. Writing can never fully represent spoken language, nor does it need to.

The idea that natural language is the primary medium of thought is commonly held. Carruthers (2002) refers to it as the standard social science model. On the account of language and cognition Carruthers (2002) presents, logical form is the interface between language and central cognitive processes, and explicit, rational thinking consists in the formation and manipulation of logical forms. That is, a person can think about, or "mentally generate," the logical form of a sentence without actually saying it. In this case, the sentence is not spoken but only thought about, or mentally represented. Once it is represented in mind, it can serve as an aid to reflective thought processes.

The function of writing in cognition may be conceived of in an analogous manner. Instead of mentally generating the logical form (LF) of a sentence, as occurs on Carruthers' (2002) account, LF is *recovered* from the permanent written representation of a sentence or linguistic expression. Writing could considerably enhance the function of LF representations in System 2 cognition because of the effects of permanence in working memory (see 2.3.5). That is, the permanence of written representations simultaneously enhances the capacity of working memory and makes available a wider range of content, including cumulative cultural archives of a culture (2.4.1), the sort represented in MUL.APIN. This exponentially expands both the type and quantity of representational content available in System 2. Writing could, in this way, amplify the role of LF in System 2 cognition and bias System 2 toward the sorts of inferences that LF underwrites.

## 7.2 *The Mind's Confrontation with Its Own Invention*

The distinctions enabled by a dual system model, detailed in 7.1, also invite the obverse consideration: how does System 1 cognition construe writing? Written signs represent an entirely new ontological domain. That is, written signs are physical objects. They are not natural kinds,[4] such as stars or animals, but rather artifacts: invented, constructed objects, like tools or water clocks. Like tools, clothing, and water clocks, their invention required intentionality and conscious reflection. But unlike other artifacts that emerged from System 2 cognition, written signs have *representational properties*. They are both *in* the world and *about* the world. That is, like art, icons, and symbols, they convey meaning. These could be called "representational kinds," although the generative power and flexibility of written signs make them unique. And although written signs are the product of intentional goals in System 2 cognition, once they exist in the world, System 1 cognition is necessarily active in their perception.

System 1 cognition recognizes the differences between artifacts and natural kinds, and attributes different causal properties to them that reflect their distinct ontological status. This recognition of kinds is rooted in basic categorization (3.3.1) and would have emerged early and non-consciously, as it allowed early man to make good decisions quickly and without conscious reflection. Recognizing that stones, trees, and deer shared properties with others of their kind allowed new instances to be understood immediately, rather than treated as unique individuals: properties and causes did not have to be discovered over and over again. This would have been particularly useful in the case of aggressive carnivorous beasts, in which case automatic, non-reflective recognition would lead to quick evasive action, and survival.[5]

System 1 cognition, then, when it first encountered written signs, would have reacted to them as it does to any new ontological domain: it would have tried to categorize them and sort out their causal properties. Central to sorting out the nature of written signs is understanding the distinction between the "sign" and the "signified" (Peirce, 1873/1991), or what the sign represents. The complex nature of the

---

[4] See 3.3 for an extended discussion of kinds of concepts, and the distinction between natural kinds and artifacts.

[5] See 2.2.3; and 7.1.

new domain of written signs, and the mind's attempt to sort it out, is reflected in the structure of the lists in the cuneiform corpus.

The impulse to categorize is a cognitive universal[6] and is similar in kind to the activities of the scientist. Von Soden (1936) called it *Ordnungswille.* Veldhuis (1997) states that the organization of the lists may represent something like a folk theory, although he also suggests a *discontinuity* between the lists and true science. Recent cognitive development work, however, comes down squarely on the side of *continuity*. Prior to examining the specific features of the cuneiform list corpus, we examine current cognitive perspectives on the development and organization of knowledge.

## 7.3 *Lists, Science, and Domains of Knowledge*

The human mind naturally seeks, not only to categorize, but also to *explain* what it sees. It recognizes that different sorts of objects require different sorts of explanations. Concepts and categories (3.3) are thus not isolated mental representations, but are organized within broader explanatory schemas that are usually co-extensive with particular domains of knowledge and the respective ontological phenomena over which those explanations function. These explanatory schemas are usually referred to as "theories," and they emerge very early in the cognitive development of individuals—so early, in fact, that they are considered to be grounded in the innate cognitive endowment of the human species. Children, it seems, have theories about the world that are similar, in significant respects, to mature scientific theories (Carey & Spelke, 1996; Wellman & Gellman, 1992). Even very young children recognize that inanimate objects move for different reasons (gravity, external physical force) than animals (drives, instincts) or people (intentions).

Naïve physics, for example, or expectations regarding characteristics and causal properties in the domain of physical objects, has been shown to emerge in infancy (Kim & Spelke, 1999). Naïve biology is well-developed by around four to six years of age (Gelman, 2003); that is, children know that although toy bears and real bears are both called "bears," the living animal has the same kinds of insides as other living kinds, such as a fox or a horse, and different from the insides of

[6] See Categories, Chapter 3.

the toy bear. Superordinate taxonomic classification, such as "animal" or "toy," thus overrides common names in children's inferential reasoning. Like Aristotle, it seems, four-year-olds are essentialists.

The mind, then, naturally sorts objects into categories (3.3). Moreover, it organizes those categories into broader systems with domain-specific, causal explanatory principles. From childhood, human knowledge is organized in coherent ways that allow us to explain the world to ourselves. Extensive empirical research in human development over the last couple of decades has amassed support for the notion of theory-based understanding in children, and it is the dominant view on the nature of early knowledge. It must be emphasized that the nature of the mature human capacity for science is still as much debated within cognitive science as it is in other disciplines but mature scientific theories can be, and currently are, viewed as *continuous* with these early, naïve theories (Carey & Spelke, 1996; Wellman & Gelman, 1992). This naturalistic, continuity perspective suggests a re-consideration of the lists in the cuneiform corpus.

### 7.4 *A Cognitive Influence on the Organization of the Lists*

The list *as a textual form* may historically precede other written forms, such as linear narratives, because the list structure reflects the mind's initial reaction to written signs. When confronted with a new ontological domain, the mind categorizes the objects within it, as if asking itself the questions: *what are those things, and what do they do*? However, the cuneiform lists seem to represent not one organizing principle, but many. Why might this be so?

Written signs were still evolving during the period in which the cuneiform corpus emerged. The environment was both multi-lingual and multi-cultural, and underwent many changes of its own during the same period. Within this dynamic context, the scribes were thrust into the role of shepherding an entirely new kind of "beast": the written sign. They were discovering the causal principles of a new ontological domain. It is not surprising that they would have wanted to bring some order and stability to the new world of signs, and the many organizational elements evident in Civil's (1995) taxonomy attest to this.

In the earliest sign lists from the Fara period, the ordering criteria are not apparent, but the "explanatory nature" of one Ebla sign list "is shown by the presence (side by side) of easily confused signs corresponding to a single sign in the primary list" (Civil, 1995:2309). In the list known as "Proto-Ea," the signs are arranged according to the stylus strokes needed to write them; one of the Late Babylonian recensions of this list was used as a didactic tool until the end of the period of active use of cuneiform writing.

The list known as:

> Proto-Izi...an Early Old Babylonian series, is organized in a sequence which is basically graphic, but with many thematic and phonological associations. For instance, one section lists eight terms for "road," with no common initial sign. It is followed by a twenty-line section with ŠID, purely acrographic, and then by phonological association with the reading sid, comes $sig_4$, "brick," and its various types.
>
> Civil, 1995:2310.

The format of the lists may thus be partially explained by the simultaneous operation of System 1, automatic, or non-reflective responses, and System 2 reflective-analytic concerns. That is, the cognitively-basic impulse to categorize objects of all sorts, including the new ontological domain of written signs, is operating at the same time as the pedagogical and/or archival concerns of System 2 cognition. Different principles of organization are intruding upon each other. The lists don't map onto modern scientific taxonomies because the scribes had yet to sort out the distinction between the *sign* and the *signified*, the "word" and its meaning, along with the properties of the written signs that were in the process of being adapted to represent diverse languages. It took millennia to sort this out.

Aristotle still hadn't sorted it out, but according to Harris (2000:90) his linguistics allowed him to proceed as if it didn't matter. Aristotle's notion of definitions and essences, that is, that they mutually determined each other, was an attempt to understand *things* rather than *words* (Robinson, 1954; cf. Harris, 2000). But Aristotle, like all literate scholars, would have been influenced by "scriptist" assumptions, to use Harris's term, or in other words, by literate conceptions of language.[7] Like four-year-olds, then, Aristotle was an essentialist, but unlike four-year-olds, he had mature literate sensibilities. These sensibilities would have included a sense of the word and its properties.

---

[7] See Watson, 1985, 1995, and Watson & Olson, 1987, for an account that suggests Aristotle was influenced by a literate concept of "word," even though he did not explicitly acknowledge this; Harris, 2009, refers to this as "scriptism," a literate bias that operates below the level of conscious awareness.

## 7.5 Listwissenschaft: *But Is It Science?*

The scientific status of the lists has been debated since Von Soden (1936) first suggested that they represented a *Listwissenschaft*, or list science. He also noted, however, that the scientific achievement represented by the lists was rather poor, although this was at least partly a consequence of his own culturally-specific understanding.

> Von Soden's analysis is problematic because it describes ancient knowledge in categories that are alien to the society of the time. Our evidence for the scribal school suggests that knowledge of the natural world did not count as academic knowledge...biological knowledge, therefore, was certainly available in this society, but biology as a scholarly discipline did not exist.
>
> Looking at the lexical corpus from the perspective of the ancient school, the classification of lexical compilations according to modern disciplinary specializations loses much of its plausibility. There is no awareness that a list of place names belongs to another realm of knowledge than a list of trees; they belong basically to one category, and that is the category of school exercises. The aim of this scholarship was not to understand nature or geography, but to understand Sumerian and Sumerian writing.
>
> Veldhuis, 2004:82

Oppenheim (1978) expressed even greater skepticism about any attempt to attach a scientific status to the lists.

> It cannot and should not be claimed, of course, that the word lists containing the names of plants, animals, or stones constitute the beginnings of botany, zoology, or mineralogy in Mesopotamia. They are not a scientific (not even a pre-scientific) achievement; rather, they result from a peculiar interaction of a genuine interest in philology (or, at any rate, lexicography) and a traditional Near Eastern concern for giving names to all things surrounding the scribe.
>
> Oppenheim, 1978:636

Von Soden's (1936) notion of *Ordnungswille* referred to System 2-like taxonomic concerns, and along with Oppenheim's (1978) observation that the lists were not commensurate with scientific taxonomies, makes far more sense on a cognitive perspective. Von Soden and Oppenheim may have been noticing the diverse, often contradictory organizational features of the lists. Their apparent difference of opinion also rests on a binary conception of "science" and "non-science," the boundaries of which are increasingly difficult to demarcate, and in any case, are not meaningful on the current cognitive perspectives reviewed in this chapter.

The lists reflect a natural impulse to categorize and explain that is both cognitively basic and continuous with mature science. They also reflect the particular context and concerns of the Mesopotamian scribes (see 1.6; 2.1). The pedagogical concerns of the scribal academy clearly had a massive influence on both the form and content, and also the survival, of the lists (Civil, 1995, 2000; Veldhuis, 1997, 2004). Administrative and economic concerns influenced the content of the earliest lists from Uruk which include those very terms needed for everyday documentation of economic activities (Nissen, Damerow, & Englund, 1993). Survival and transmission of the lists can also be attributed to strong cultural traditions, such as reverence for received knowledge and for writing itself (Black & Tait, 1995; Pearce, 1995; Veldhuis, 2004). Many of the Archaic Uruk lists, like those in the later Urra = *hubullu* tradition, are thematically organized, and thus may be used either for scribal education or as a reference tool. In other words, they serve an encyclopedic function.

While categorization is a basic property of mind, then, and a characteristic of System 1 cognition, the impulse to write a list necessarily engages System 2 reflective-analytic cognition. Whether the content of the list is stars, constellations, planets or other objects, most of the lists in the cuneiform corpus are long enough to suggest that they exceed the capacity of ordinary working memory.[8] On capacity considerations alone, then, it seems clear that writing played a constitutive role in the appearance of lists. In addition to capacity considerations, the content and organization of the lists also mark them unmistakably as products of a literate sensibility (Civil, 1995, 2000; Veldhuis, 1997, 2004). A naturalistic perspective suggests that further influences may be operating, and may shed further light on the nature of the star lists with which MUL.APIN begins, and the development of categorical expressions through the text.

## 7.6 *Star Lists and the Extended Function of Writing in MUL.APIN*

With respect to the astronomical content of MUL.APIN, the star lists represent the *set over which subsequent generalizations are made*. That is, the stars are first *named* in list entries, then *grouped* into categories by

[8] See Miller, 1956; and Baddely, 2007, for working memory capacity.

summary statements, then followed by the description of parameters governing their visibility and trajectories, in the general progression toward the complex procedures that appear later in the text. On the above account, however, the star lists also reflect the basic organizational properties of the human mind, and centuries of analysis and reflection, as well as scribal traditions and conventions.

Conceptually, the star lists represent a category of natural kinds grounded in basic cognition, a category which assumes new significance when organized and set down in written form. When considered within the MUL.APIN treatise as a whole, the star lists can serve to support new inferences in the kind of reflection and analysis characteristic of System 2 cognitive processes. That is, when they are considered together with subsequent component sections of the treatise, they allow new inferences to be made.

On a dual-system model of cognition (7.1, above) written content can be expected to serve at least two functions in System 2 cognition. First, writing boosts "on-line" working memory capacity, enlarging the amount of information which can be simultaneously considered. Second, the archival properties of written records extend the cumulative body of knowledge that an individual can bring to bear. The content of reflection and analysis is not restricted to what an individual can think about, or mentally generate, based on personal knowledge and experience. Rather, it expands to include the *cumulative record of the culture*.

Writing, then, gives a boost to working memory, and also adds to the sum total of available, explicitly represented knowledge. This amplification, or enhancement, of System 2 cognitive abilities is not neutral. To the extent that writing isolates the logical form of language (see 2.4.4; and 7.1, above) it biases cognitive processes toward the sorts of inferences that LF underwrites, which in turn could be expected to bias thought toward logical, rational thought.

It may also be the case that this isolation of logical form also recruits the property of *generativity*.[9] In other words, logical form representations bring the *combinatorial properties of syntax* into the conscious mind. Fodor (1994, 2001) argues that thought operates in a manner that

[9] Generativity refers to the property of human language, whereby a finite set of grammatical rules can *generate* an infinite number of meaningful sentences; Fodor, 2001, and discussion in 3.5.1.

parallels the componential, generative properties of syntax. The mind can exploit combinatorial properties to generate novel, abstract conceptual categories (see 3.5.1). The permanent, external representation of linguistic form in writing, to the degree that it enables the recovery of logical form, could conceivably enhance this process. That is, the mind could exploit cumulative, archival representations to an even greater effect. Writing, in this way, could "bootstrap" the uniquely-human propensities toward the abstract, creative thought necessary for the development of a formal theoretical science. Science, like syntax, is both infinitely creative and uniquely human.

We can now more fully appreciate the influence, on the astronomer-scribes, of the particular forms of expression that appear in MUL. APIN at II ii 7 *You lo[ok(?)] for the risings (?) and... of the stars of Ea, Anu and Enlil*, and the axiom at II iii 13 *4 is the coefficient for the visibility of the Moon.* The expressions, and any inferences they afford, can serve as input to reflective-analytic thought within System 2 cognition. At the same time, all of the information represented in the foregoing component sections of the treatise is simultaneously accessible. The componential, cumulative nature of the treatise, then, provides a rich inferential environment within which to interpret the statements at II ii 7 and II iii 13. For this reason, the two statements could be expected to yield different interpretive results, and therefore different cognitive effects, than they would if read within the texts of either the omen series or the coefficient lists.

Further, in the case of the axiom at II iii 13, the *linguistic form itself* could serve as the basis for inference. That is, the axiom has the form of an abstract statement of equivalence. By way of simple substitution of abstract terms, it could "bootstrap" the inductive leap to a purely formal, stipulative definition, the primary building block of a theoretical science. Writing, then, a *cultural* invention, amplifies the *natural* properties of mind. It does this in very simple, straightforward way, under the pressure of effortful extended inferential processing.

## 7.7 *Summary*

On naturalistic accounts of the development of knowledge, scientific theories develop out of the core conceptual knowledge common to every human being. Astronomical science, of the observational variety found in MUL.APIN, begins with the taxonomies represented by the star lists, then develops through the forms and functions identified and discussed in Chapters 4 through 6.

Writing may figure in this process in a number of ways. The permanence of writing renders content available over time for effortful, conscious, analytic processes. It boosts the capacity of working memory, and it extends access beyond information stored in individual memory to that recorded in the cumulative archival records of the culture. Writing down the knowledge represented in the star lists, and the simultaneous recording of generalizations, measurements, and procedures in a single treatise, would allow the astronomer scribes to consider the entire body of available astronomical knowledge at one time. By boosting System 2 cognitive processes in this manner, writing conceivably, and quite probably, "bootstrapped" the development of formal astronomical science.

CHAPTER EIGHT

# A FINAL WORD: FROM LIST TO AXIOM

The claim that MUL.APIN represents a significant milestone in the development of Astronomical science requires reconciliation with at least two strands of argument. First is the relation of the text to the broader context of the centuries-old Mesopotamian scribal tradition. While MUL.APIN is a unique text in the cuneiform corpus, without direct written precursors, it bears similarity to a particular strand of that tradition popular at the time and place that MUL.APIN first appears on clay, and that is the "technical handbooks," or "procedural texts." These occur widely in Neo-Assyrian times, particularly in Assurbanipal's library.

A second consideration is the presence of anomalous text in MUL.APIN that, to the modern mind at least, seems to place it squarely in the category of "non-science." The presence of omens in section m might appear to relegate the entire text to the realm of superstition. We address each of these concerns in turn.

## 8.1 *MUL.APIN and the Technical Handbook Tradition*

At least two features of MUL.APIN seem to place it within the technical handbook tradition. First, the inclusion of *technical apparatus* in MUL.APIN, such as the gnomon and the water clock used to measure time, bears substantial similarities with other procedural texts. The second feature is the use of *second person direct address* of the reader of the text. Both of these features seem to place MUL.APIN within the technical handbook tradition.

The best preserved and best known example of the technical handbooks may be the glass texts published by A. L. Oppenheim in *Glass and Glassmaking in Ancient Mesopotamia* (Moorey, 1999; Oppenheim 1970). Here, the author/speaker addresses the reader in second-person form with detailed instructions as to how to make various types of glass. A typical set of instructions is as follows:

> If you want to produce *zagindurû*-colored glass, you grind finely, separately, 10 minas of *immanakku*-stone, 15 minas of naga-plant ashes (and) 1 2/3 minas of "White Plant." You mix these together. You put (them) into a cold kiln which has four fire openings and arrange the mixture between the openings. You keep a good and smokeless fire burning until the "metal" (molten glass) becomes fritted. You take it out and allow it to cool off.
>
> Oppenheim, 1970:35

Mesopotamian mathematical texts take a similar format. Instructions, again given in the second person, provide the procedure for solving various types of set problems and equations, or explain how to make certain measurements in the case of geometric shapes.

> 1 cubit is the circumference of a log (cylinder). How thick was it? You, multiply 5/60 × 5/60. 25/3,600 you will multiply by 4, 48, the fixed coefficient (with the result of) 2. 2 sìla is the thickness of the log.
>
> Adapted from Neugebauer, 1945:57–58

Similar texts in this 'How to make/do...' format are extant for making metal alloys (Oppenheim, 1966), cooking recipes (Bottero,1995), making perfume, brewing beer, practicing medicine, performing rituals, and taking omens (Oppenheim,1970:5–7; Moorey, 1999:6).

The use of second person address in MUL.APIN begins in section e and is found in succeeding sections. When considered together with the technical-procedural content of the text, it suggests that MUL. APIN can be placed within the technical handbook tradition. In this sense, then, MUL.APIN can be considered a manual, a handbook, or a textbook, that might be called "How to practice astronomy using the accepted techniques of the traditional astronomy of the Neo-Assyrian royal court." MUL.APIN provides the methods and procedures which underlie the raw observations and astrological interpretations of the Neo-Assyrian royal astronomers, which are preserved for us in letters and reports to the Assyrian king (SAA 8, Hunger, 1992; and SAA 10, Parpola, 1993).

The latter sections of MUL.APIN may even serve as a bridge to the later procedural texts in the ACT tradition of the Persian and Hellenistic periods. These later texts maintain the second-person form, but give much more detailed instructions, with the highly exact and complex mathematical formulae required for calculations found in late-Babylonian mathematical astronomy (Neugebauer, 1955; Hunger & Pingree, 1999:183–270). It therefore may be more than coincidence that the earliest examples of the later ACT-type tradition (some of the so-called "atypical" texts) begin to appear not long after MUL.APIN (Hunger & Pingree, 1999).

In some sense, the ACT texts could even be considered more advanced versions of the handbook/textbook form, i.e. manuals for "How to do mathematical astronomy," in which the use of technical apparatus for making observations is replaced by complex mathematical systems. However, it must be emphasized that the ACT texts represent an exponential increase in sophistication over earlier astronomical and procedural texts, including a unified and comprehensive lunar theory with elegant mathematical models for the prediction of the effects of lunar and solar anomalies, the recent examination of which has attested to their accuracy and predictive power (Britton, 2007, 2009).

The sequence of sections in MUL.APIN, on our view, thus parallels the history of the development of ancient Mesopotamian astronomical texts, starting with lists and ending with procedural instructions. Just as the succeeding sections of the MUL.APIN treatise offer increasingly complex instructions, they similarly require more sophisticated astronomical knowledge, reflecting ongoing improvements in astronomical technique. The progression culminates with introduction of the technical apparatus, the gnomon and the water clock, used to measure time, and also reflects a progression from lists to procedures.

MUL.APIN represents the cumulative body of astronomical knowledge in the Mesopotamian cuneiform tradition. As far as extant records allow us to assume, it represents the first time this cumulative knowledge could be simultaneously, and repeatedly, considered in a systematic manner. On the naturalistic perspective outlined in the foregoing chapter (7), it is entirely conceivable, even plausible, that MUL.APIN was a cornerstone in the development of the sophisticated astronomical science represented in the ACT tradition. Thus, it may have played a role as pivotal in ancient Mesopotamian astronomy as that played by Copernicus or Galileo in modern astronomy. On any account it appears to be a transitional text, a bridge from older forms of thought, such as the omens, to the early phases of true mathematical astronomy. We now consider it in relation to the omen tradition, which continued to be transmitted in writing to the end of cuneiform writing.

## 8.2 *The Omens and Anomalous Text*

Most of the component sections of MUL.APIN refer to observations and calculations of the type we associate with a scientific tradition. But there remain segments of the treatise which seem to defy definition as "scientific," at least in modern terms. How, then, can we suggest that the text is "scientific" if it includes this type of material? We address this first historically, and then from current perspectives on science and logic as rational activities.

First, it is paramount to view the omen and anomalous text material from an ancient Mesopotamian perspective. The omen tradition, and astrology in general, was considered to be a rational and valid form of knowledge until well into the period when late mathematical astronomy developed and flourished. In fact, the new astrology of the zodiac and horoscope only emerges in this late period, contemporaneously with the highly sophisticated ACT tradition.[1] Further, the durability of the Enuma Anu Enlil omen tradition, the origins of which substantially pre-date MUL.APIN, must be understood as reflecting the veneration of older, received, knowledge by first-millennium Mesopotamian astronomer-scribes.

The placement of the omens at the very end of MUL.APIN may signify that the activities of the professional astronomers in MUL.APIN were relevant to the everyday concerns of Assyrians and Babylonians.[2] The Mesopotamians regarded observable heavenly phenomena as the "writing" of the gods, information revealed to mankind for use in understanding lived events (Rochberg, 2004). Our doubts regarding the rational nature of the omens reflect anachronistic assumptions: the presence of horoscopes in most modern newspapers is not usually taken as evidence that the newspaper editors who approve their publication *believe* in them. The presence of omens in MUL.APIN attests to their cultural significance, not necessarily to the underlying beliefs of the scribes.

It is also the case that omens have a predictive structure: If X happens (protasis), Y is the consequence (apodosis); if X happened again, one could expect Y to happen again. The linguistic form of the omens is conditional: *If p, then q.* Rochberg (in press; cf. Rochberg-Halton, 1982)

---

[1] See footnote 5 in the introduction to this volume.

[2] See discussion in Chapter 4, section 4.11; particularly section 4.11.4.1.

suggests that this conditional, predictive structure identifies omens as belonging to the same sort of rational activity we associate with science. They are *rational with respect to the belief system in which they operate.* It is the causal-explanatory principle underlying them ("the gods") that differs from the assumptions of modern scientist, not their inherent rationality:

> The analysis of the conditional form of Babylonian omens shows that though the omen statements certainly posit relations between phenomena that do not depend upon the physical and causal connections we ourselves would make, the relation between protasis and apodosis is a logically valid one that furthermore can be classified with inferences expressed in the form of conditionals. Inferential reasoning... thereby lies at the basis of the connections between the propositions of antecedent and consequent.
>
> Rochberg (in press)

Omens are not the products of irrational minds. They differ from modern science, but are nevertheless rational with respect to the belief system in which they occur. The predictive structure of the omens, however, is clearly not commensurate with predictive power: modern-day physics makes substantially better predictions. The astronomer scribes who composed MUL.APIN were able to write an axiom pertaining to the visibility of the moon, but they were a long way from being able to land on it. Their science was not that far advanced.

We can offer no account comparable to that of the omens for text of the sort found in subsection h-i-1 (see Chapter 4, 4.7.1), which is concerned with wind and weather. It is difficult for the modern reader to construe this text component in any meaningful way, but we assume that this material had meaning to the authors of MUL.APIN and its readers. Otherwise, it would not have survived in the canonical text.

## 8.3 *MUL.APIN, Science, and Rationality*

If science develops out of core cognition, common to all human beings (7.3), understanding its origins and development may involve a deeper exploration of the nature of System 2 cognition (7.1) and of rationality in general (cf. Harris, 2005, 2009). There seems little basis for claiming that the impulse behind the lists in the cuneiform corpus is different in *kind* from the impulse that underlies, for example, Linnaean taxonomies, even though the end results are widely divergent.

Cognitive-developmental evidence would suggest that both are continuous with innate properties of the human mind (7.3).

Yet to the modern mind, much of the "logic" in the cuneiform corpus remains inchoate. Rationality, on Harris's (2009:107) account, involves grasping a network of actual and potential beliefs, and it hinges on the sign-making capacity that human beings exercise in their communication and social organization. The *Listwissenschaft* of the cuneiform corpus is clearly rational, a culturally sanctioned version of a universal impulse that developed cumulatively, over centuries and millennia, and that was transmitted within a cultural tradition. The observational science of MUL.APIN appears to occupy a pivotal role in the development of the later, more sophisticated mathematical-astronomy of the ACT tradition. However, in the absence of textual evidence, this claim must remain speculative.

Whatever factors underlie the exponential advance into theory represented by the emergence of System A and B Babylonian mathematical astronomy,[3] as it appears in the ACT tradition, it seems likely to have been enabled by a cumulative text-based tradition, probably collaborative, multi-cultural, and of many centuries duration. Similarly, the Hellenistic transmission of this tradition[4] is highly likely to have influenced later thought.

The identification of rationality and of logic with the Greeks, and with Aristotle in particular, is deeply embedded in the Western intellectual tradition. The intellectual achievements of the classical period remain as compelling today as they were to contemporary scholars. Yet the Greeks saw clear and far, in no small part, because they stood on the shoulders of others, as modern scientists stand on theirs.

Science is a cumulative enterprise, rooted in the basic human impulse to observe and explain. On our view, MUL.APIN was a major stepping stone along the path. It presents the cumulative record of a foundational observational science that emerged within the Mesopotamian text tradition. It begins with one of the simplest extant written forms—the list—and ends with one of the most formal and abstract, the definition of an axiomatic concept. We have argued that the relative placement of these written forms is not coincidental, and that the treatise does not represent a random collection of related texts. On

[3] See Britton, 2007, 2009; for the ACT tradition, see Chapter 1, fn. 5.
[4] See Chapter 2, 2.3.2.

our view, the treatise reflects a clearly ordered sequence that parallels the development of the text tradition, with earlier text forms appearing earlier in the MUL.APIN treatise. Even in the absence of definitive proof of an historical sequence, the treatise clearly reflects a developmental progression in both astronomical *content* and textual *form*. The changes are more consistent with a dynamic process of conceptual change than with a static model of accretion.

We can't know for certain the extent to which the ancient compilers, editors, and readers of the MUL.APIN treatise were fully cognizant of the developmental progression in which they appear to have been engaged. Nevertheless, the evidence for recalibration, such as rhetorical-indexical clusters and the use of the marker DIŠ, suggest that the scribes knew what they were doing, and why and how they were doing it, even though the convention of explaining their decision-making processes in writing was outside the experience of their time and place.

MUL.APIN may not represent fully developed science, but it does offer a unique, even vital, window onto its beginnings, and the dynamic, reflective processes involved in the emergence of a formal written science. The transition from star lists at the beginning of the treatise to a formally expressed axiom near its end represents a quantum leap in both conceptual content and textual form. This axiom, *4 is the coefficient for the visibility of the moon*, puts the astronomer scribes who wrote it well within reach of a formal, theoretical, mathematical science.

# BIBLIOGRAPHY


Alegria, J., E. Pignot, & J. Morais. 1982. "Phonetic Analysis of Speech and Memory Codes in Beginning Readers." *Memory and Cognition* 10:451–456.

Al-Rawi, F., & W. Horowitz. 2001. "Tablets from the Sippar Library IX: A *Ziqpu*-Star Planisphere." *Iraq* 63:171–181.

Antony, L. M., & N. Hornstein. 2003. *Chomsky and His Critics*. London: Blackwell.

Atkinson, D. 1999. *Scientific Discourse in Sociohistorical Context*. Mahwah, NJ: Earlbaum.

Atran, S. 1990. *Cognitive Foundations of Natural History*. Cambridge: Cambridge University Press.

Atran, S., D. Medin, & N. Ross. 2005. "The Cultural Mind: Environmental Decision Making and Cultural Modeling within and across Populations." *Psychological Review* 112:744–776.

Baddeley, A. 2007. *Working Memory, Thought, and Action*. Oxford: Oxford University Press.

Bar-Hillel, Y. 1954. "Indexical Expressions." *Mind* 63:359–379.

Bertelson, P., & B. de Gelder. 1994. "The Cognitive Psychology of Literacy: Some Basic Findings." Pp. 151–162 in *Functional Literacy: Theoretical Issues and Educational Implications*, edited by L. Verhoeven. Amsterdam: Benjamins.

Biebuyck, D., & K. C. Mateene. 1971. *The Mwindo Epic*. Berkeley: University of California Press.

Bierwisch, M., & F. Kiefer. 1969. "Remarks on Definitions in Natural Languages." Pp. 55–79 in *Studies in Syntax & Semantics*, edited by F. Kiefer. Dordrecht: Reidel.

Black, J., & W. J. Tait. 1995. "Archives and Libraries in the Ancient Near East." Pp. 2197–2201 in *Civilizations of the Ancient Near East*, edited by J. Sasson. New York: Charles Scribner's Sons.

Boroditsky, L. 2001. "Does Language Shape Thought? Mandarin and English Speakers' Conception of Time." *Cognitive Psychology* 43:1–22.

Bottero, J. 1995. *Textes culinaires mésopotamiens (Mesopotamian Culinary Texts)*. Winona Lake, IN: Eisenbrauns.

Bowerman, M., & S. C. Levinson. 2001. *Language Acquisition and Conceptual Development*. Cambridge: Cambridge University Press.

Boyd, R., & P. J. Richerson. 1985. *Culture and the Evolutionary Process*. Chicago: University of Chicago Press.

Brack-Bernsen, L. 2007. "The 360-Day Year in Mesopotamia." Pp. 83–100 in *Calendars and Years: Astronomy and Time in the Ancient Near East*, edited by J. Steele. Oxford: Oxbow Books.

Bradley, L., & P. E. Bryant. 1983. "Categorizing Sounds and Learning to Read: A Causal Connection." *Nature* 301:419–421.

Breheny, R. 2002. "The Current State of (Radical) Pragmatics in the Cognitive Sciences," in "Pragmatics and Cognitive Science," special issue, *Mind and Language* 17, nos. 1–2:169–187.

Britton, J. 2002. "Treatments of Annual Phenomena in Cuneiform Sources." Pp. 21–78 in *Under One Sky: Astronomy and Mathematics in the Ancient Near East*, edited by J. Steele and A. Imhausen. Münster: Ugarit-Verlag.

——. 2007. "Studies in Babylonian Lunar Theory: Part I. Empirical Elements for Modeling Lunar and Solar Anomalies." *Archive for History of Exact Sciences* 61:83–145.

——. 2009. "Studies in Babylonian Lunar Theory: Part II. Treatments of Lunar Anomaly." *Archive for History of Exact Sciences* 63:357–431.

Brown, D. 2000. *Mesopotamian Planetary Astronomy-Astrology*. Groningen: Styx Publications.
Brown, D., J. Fermor, & C. B. F. Walker. 1999. "The Water Clock in Mesopotamia." *Archiv für Orientforschung* 46:130–148.
Brown, R. 1958. "How Shall a Thing Be Called?" *Psychological Review* 65:14–21.
Brug, J. F. 1990. "Biblical Acrostics and Their Relationship to Other Ancient Near Eastern Acrostics." Pp. 283–304 in *The Bible in the Light of Cuneiform Literature: Scripture in Context III*, edited by W. W. Hallo, B. W. Jones, and G. L. Mattingly. Ancient Near Eastern Texts and Studies 8. Lewiston, NY: Edwin Mellen Press.
Carey, S. 1985. *Conceptual Change in Childhood*. Bradford Books. Cambridge, MA: MIT Press.
——. 2000. "Science Education as Conceptual Change." *Journal of Applied Developmental Psychology* 21:13–19.
——. 2001. "Evolutionary and Ontogenetic Foundations of Arithmetic." *Mind and Language* 16:37–55.
Carey, S., & E. Spelke. 1996. "Science and Core Knowledge." *Philosophy of Science* 63:515–533.
Carnap, R. 1937. *The Logical Syntax of Language*. Translated by A. Smeaton. London: Kegan Paul.
Carruthers, P. 2002. "The Cognitive Functions of Language." *Behavioral and Brain Sciences* 25:657–726.
——. 2006. *The Architecture of Mind: Massive Modularity and the Flexibility of Thought.* Oxford: Oxford University Press.
Carston, R. 2002. "Linguistic Meaning, Communicated Meaning, and Cognitive Pragmatics," in "Pragmatics and Cognitive Science," special issue, *Mind and Language* 17, nos. 1–2:127–148.
Casaburi, M. C. 2003. *Tre-stella-per-ciascun(-mese): L'Astrolabio B, edizione filologia*. Naples: Università degli studi di Napoli "L'Orientale."
Castles, A., & M. Coltheart. 2004. "Is There a Causal Link from Phonological Awareness to Success in Learning to Read?" *Cognition* 91:77–111.
Castro-Caldas, A., K. M. Peterson, A. Reis, S. Stone-Elander, & M. Ingvar. 1998. "The Illiterate Brain: Learning to Read and Write during Childhood Influences the Functional Organization of the Adult Brain." *Brain* 121:1053–1063.
Chemla, K. 2004. *History of Science, History of Text*. Boston: Springer.
Cheung, H., H. Chen, C. Lai, O. Wong, and M. Hills. 2001. "The Development of Phonological Awareness: Effects of Spoken Language Experience and Orthography." *Cognition* 81:227–241.
Civil, M. 1995. "Ancient Mesopotamian Lexicography." Pp. 2305–2314 in *Civilizations of the Ancient Near East*, edited by J. Sasson. New York: Charles Scribner's Sons.
——. 2000. "From the Epistolary of the Edduba." Pp. 105–118 in *Wisdom, Gods, and Literature: Studies in Assyriology in Honor of W. G. Lambert*, edited by A. R. George and I. L. Finkel. Winona Lake, IN: Eisenbrauns.
Clark, H. 1996. *Uses of Language*. Cambridge: Cambridge University Press.
Cooper, J. S. 2004. "Babylonian Beginnings: The Origin of the Cuneiform Writing System in Comparative Perspective." Pp. 71–99 in *The First Writing: Script Invention as Theory and Process*, edited by S. D. Houston. Cambridge: Cambridge University Press.
Daniels, P. T., & W. Bright. 1996. *The World's Writing Systems*. New York: Oxford University Press.
Dascal, M. 1987. *Leibniz: Language, Signs, and Thought*. Amsterdam: John Benjamins.
Davidson, D. 1990. "The Structure and Content of Truth." *The Journal of Philosophy* 87:279–328.
Dawkins, R. 1976. *The Selfish Gene*. New York: Oxford University Press.
——. 1982. *The Extended Phenotype*. Oxford: Oxford University Press.
Dehaene, S., V. Izard, P. Pica, & E. S. Spelke. 2006. "Core Knowledge of Geometry in an Amazonian Indigene Group." *Science* 311:381–384.

Dennett, D. 1994. "Language and Intelligence." Pp. 161–178 in *What Is Intelligence?*, edited by J. Khalfa. Cambridge: Cambridge University Press.

——. 1998. "Reflections on Language and Mind." Pp. 284–294 in *Language and Thought*, edited by P. Carruthers & J. Boucher. Cambridge: Cambridge University Press.

——. 1999. "The Evolution of Culture." The Charles Simonyi Lecture, Oxford University. http://www.edge.org/3rd_culture/dennett/dennett_p1.html.

Dienes, Z., & J. Perner. 1999. "A Theory of Implicit and Explicit Knowledge." *Behavioral & Brain Sciences* 22:735–808.

Disalle, R. 2002. "Newton's Philosophical Analysis of Space and Time." Pp. 33–56 in *The Cambridge Companion to Newton*, edited by I. B. Cohen & G. E. Smith. Cambridge: Cambridge University Press.

Englund, R. K. 1988. "Administrative Timekeeping in Ancient Mesopotamia." *Journal of the Economic and Social History of the Orient* 31:121–185.

——. 1998. "Texts from the Late Uruk Period." Pp. 15–233 in *Mesopotamien: Späturuk-Zeit und frühdynastiche Zeit*, edited by J. Bauer, R. Englund, & M. Krebernik. Orbis Biblicus et Orientalis 160/1. Freiberg: Universitätsverlag; Göttingen: Vandenhoek & Ruprecht.

Evans, J. St. B. T., & K. Frankish. 2009. *In Two Minds: Dual Process and Beyond*. Oxford: Oxford University Press.

Evans, J. St. B. T., & D. E. Over. 1999. "Explicit Representation in Hypothetical Thinking." *Behavioral & Brain Sciences* 22:763–764.

——. (2004) *If*. Oxford: Oxford University Press.

Fillmore, C. 1997. *Lectures on Deixis*. Stanford, CA: CSLI Publications.

Fitch, W. T. 2007. "Bio-Linguistics: The Biology and Evolution of Language." Plenary address presented at the annual conference of the International Linguistics Association, New York, March 31.

Fitch, W. T., & Reby, D. 2001. "The Descended Larynx Is Not Uniquely Human." *Proceedings of the Royal Society, Biological Sciences* 268:1669–1675.

Fodor, J. A. 1983. *The Modularity of Mind: An Essay in Faculty Psychology*. Cambridge, MA: MIT Press.

——. 1994. "Concepts: A Potboiler." *Cognition* 50:95–113.

——. 2001. "Language, Thought, and Compositionality." *Mind & Language* 16:1–15.

Fox, D. 2003. "On Logical Form." Pp. 82–123 in *Minimalist Syntax*, edited by R. Hendrick. Oxford: Blackwell.

Frankish, K. 2009. "Systems and Levels: Dual-System Theories and the Personal-Subpersonal Distinction." Pp. 89–107 in *In Two Minds: Dual Process and Beyond*, edited by J. Evans & K. Frankish. Oxford: Oxford University Press.

Frege, G. 1892. "Über Sinn und Bedeutung." *Zeitschrift für Philosophie und philosophische Kritik* 100:25–50.

Frith, U. 1998. "Literally Changing the Brain." *Brain* 121:1011–1012.

Gaur, A. 1987. *A History of Writing*. London: The British Library.

Gelb, I. J. 1963. *A Study of Writing*. 2nd ed. Chicago: University of Chicago Press.

Geller, M. 1997. "The Last Wedge." *Zeitschrift für Assyriologie* 87:43–96.

Gelman, S. 2003. *The Essential Child: Origins of Essentialism in Everyday Thought*. New York: Oxford University Press.

——. 2004. "Psychological Essentialism in Children." *Trends in Cognitive Science* 8:404–409.

Gentner, D. 2003. "Why We're So Smart." Pp. 195–235 in *Language in Mind: Advances in the Study of Language and Thought*, edited by D. Gentner & S. Goldin-Meadow. Cambridge, MA: MIT Press.

Gentner, D., & S. Goldin-Meadow. 2003. *Language in Mind: Advances in the Study of Language and Thought*. Cambridge, MA: MIT Press.

George, A. R. 1991. Review of H. Hunger & D. Pingree, *MUL.APIN: An Astronomical Compendium in Cuneiform*. *Zeitschrift fur Assyriologie* 81:301–306.

——. 2003. *The Babylonian Gilgamesh Epic: Introduction, Critical Edition, and Cuneiform Texts.* Oxford: Oxford University Press.
Gladwin, T. 1970. *East Is a Big Bird: Navigation and Logic on a Puluwat Atoll.* Cambridge, MA: Harvard University Press.
Goody, J. 1977. *The Domestication of the Savage Mind.* Cambridge: Cambridge University Press.
——. 1987. *The Interface between the Written and the Oral.* Cambridge: Cambridge University Press.
Goody, J., & I. Watt. 1968. "The Consequences of Literacy." Pp. 27–68 in *Literacy in Traditional Societies*, edited by J. Goody. Cambridge: Cambridge University Press. Originally published in *Contemporary Studies in Society and History* 5 (1963): 304–345.
Grice, H. P. 1968. "Utterer's Meaning, Sentence Meaning, and Word Meaning." *Foundations of Language* 4:225–242.
——. 1975. "Logic and Conversation." Pp. 41–58 in *Syntax and Semantics*, vol. 3, edited by P. Cole & J. L. Morgan. New York: Academic Press.
——. 1989. *Studies in the Way of Words.* Cambridge, MA: Harvard University Press.
Gross, A., J. Harmon, & M. Reidy. 2002. *Communicating Science: The Scientific Article from the 17th Century to the Present.* Oxford: Oxford University Press.
Gunter, A. 2009. *Greek Art and the Orient.* Cambridge: Cambridge University Press.
Halverson, J. 1992. "Goody and the Implosion of the Literacy Thesis." *Man* 27:301–317.
Harman, G. 1996. "Analyticity Regained?" *NOÛS* 30:392–400.
Harris, R. 1986. *The Origin of Writing.* London: Duckworth.
——. 2000. *Rethinking Writing.* London: Athlone Press.
——. 2005. *The Semantics of Science.* London: Continuum.
——. 2009. *Rationality and the Literate Mind.* New York: Routledge.
Harris, W. V. 1989. *Ancient Literacy.* Cambridge: Cambridge University Press.
Hauser, M., N. Chomsky, & W. T. Fitch. 2002. "The Faculty of Language: What Is It, Who Has It, and How Did It Evolve?" *Science* 298:1569–1579.
Havelock, E. 1963. *Preface to Plato.* Cambridge, MA: Harvard University Press.
——. 1982. *The Literate Revolution in Greece and Its Cultural Consequences.* Princeton, NJ: Princeton University Press.
Holden, C. 1998. "No Last Word on Language Origins." *Science* 282:1455–1458.
Horowitz, W. 1991. "Antiochus I, Esagil, and a Celebration of the Ritual for Renovation of Temples." *Revue d'Assyrologie* 85:75–77.
——. 1994. "Two New Ziqpu-Star Texts." *Journal of Cuneiform Studies* 43:89–98.
——. 1996. "The 360 and 364 Day Year in Ancient Mesopotamia." *Journal of the Ancient Near Eastern Society* 24:35–41.
——. 1998. *Mesopotamian Cosmic Geography.* Winona Lake, IN: Eisenbrauns.
——. 2005. "Some Thoughts on Sumerian Star-Names and Sumerian Astronomy." Pp. 163–178 in *An Experienced Scribe Who Neglects Nothing: Ancient Near Eastern Studies in Honor of Jacob Klein*, edited by Y. Sefati, P. Artzi, C. Cohen, B. Eichler, and V. Hurowitz. Bethesda: CDL Press.
——. 2007. "The Astrolabes: Astronomy, Theology, and Chronology." Pp. 101–113 in *Calendars and Years: Astronomy and Time in the Ancient Near East*, ed. J. Steele. Oxford: Oxbow Books.
——. Forthcoming. *The Three Stars Each: The Astrolabes and Related Texts.*
Horowitz, W., & J. Oelsner. 1997. "The 30 Star-Catalogue HS 1897 and the Late Parallel BM 55502." *Archiv für Orientforschung* 44/45:176–185.
Hunger, H. 1968. *Babylonische und assyrische Kolophone.* Neukirchen-Vluyn: Butzon & Bercker Kevelaer.
——. 1992. *Astrological Reports to Assyrian Kings.* State Archives of Assyria 8. Helsinki: Helsinki University Press.
Hunger, H., and D. Pingree. 1989. *MUL.APIN: An Astronomical Compendium in Cuneiform.* Archiv für Orientforschung Beiheft 24. Horn, Austria: F. Berger.

——. 1999. *Astral Sciences in Mesopotamia.* Leiden: Brill.
Hurowitz, V. A. 2002. "Additional Elements of Alphabetical Thinking in Psalm XXXIV." *Vetus Testamentum* 52:326–333.
Innis, H. A. 1999. *The Bias of Communication.* Toronto: University of Toronto Press. First published 1951.
Jolley, N., ed. 1994. *The Cambridge Companion to Leibniz.* Cambridge: Cambridge University Press.
Jones, A. 1999. *Astronomical Papyri from Oxyrhynchus (P. Oxy. 4133–4300a).* 2 vols. in 1. Memoirs of the American Philosophical Society 233. Philadelphia: American Philosophical Society.
Kant, I. 1781/1990. *Critique of Pure Reason.* Translated by J. D. M. Meiklejohn. Amherst, NY: Prometheus Books.
Kay, P. 1999. "Color Categorization." Pp. 143–145 in *The MIT Encyclopedia of the Cognitive Sciences*, edited by R. Wilson & F. Keil. Cambridge, MA: MIT Press.
Kay, P., & T. Regier. 2006. "Language, Thought, and Color: Recent Developments." *Trends in Cognitive Science* 10:51–54.
Kim, K., & E. Spelke. 1999. "Perception and Understanding of Effects of Gravity and Inertia on Object Motion." *Developmental Science* 2:339–362.
Kintsch, W. 1998. *Comprehension: A Paradigm for Cognition.* Cambridge: Cambridge University Press.
Kinzler, K. D., & E. S. Spelke. 2007. "Core Systems in Human Cognition." *Progress in Brain Research* 164:257–264.
Kneale, W., & M. Kneale. 1984. *The Development of Logic.* Oxford: Clarendon.
Koch-Westenholz, U. 1995. *Mesopotamian Astrology: An Introduction to Babylonian and Assyrian Celestial Divination.* Copenhagen: The Carsten Niebuhr Insititute of Near Eastern Studies.
Kornblith, H. 1994. *Naturalizing Epistemology.* 2nd ed. Cambridge, MA: MIT Press.
Kripke, S. 1972. "Naming and Necessity." Pp. 253–355, 763–769 in *Semantics of Natural Language*, edited by D. Davidson and G. Harman. Dordrecht: Reidel.
Kuczaj, S. A., & L. E. Highfill. 2005. "Dolphin Play: Evidence for Cooperation and Culture?" *Behavioral and Brain Sciences* 28:705–706.
Lambert, W. G. 1960. *Babylonian Wisdom Literature.* Oxford: Clarendon Press.
——. 1962. "A Catalogue of Texts and Authors." *Journal of Cuneiform Studies* 16:59–77.
Levinson, S. C. 1983. *Pragmatics.* Cambridge: Cambridge University Press.
——. 1996. "Language and Space." *Annual Review of Anthropology* 25:353–382.
——. 2003. *Space in Language and Cognition.* Cambridge: Cambridge University Press.
Li, P., & L. Gleitman. 2002. "Turning the Tables: Language and Spatial Reasoning." *Cognition* 83:265–294.
Lloyd, G. E. R. 1979. *Magic, Reason, and Experience: Studies in the Origin and Development of Greek Science.* Cambridge: Cambridge University Press.
——. 1990. *Demystifying Mentalities.* Cambridge: Cambridge University Press.
——. 2003. "Literacy in Greek and Chinese Science: Some Comparative Issues." Pp. 122–138 in *Written Texts and the Rise of Literate Culture in Ancient Greece*, ed. H. Yunis. Cambridge: Cambridge University Press.
Lloyd, P., & C. Fernyhough, eds. 1999. *Lev Vygotsky: Critical Assessments.* 4 vols. London: Routledge.
Luria, A. R. 1976. *Cognitive Development: Its Cultural and Social Foundations.* Cambridge: Cambridge University Press.
Majid, A., M. Bowerman, S. Kita, D. Haun, & S. C. Levinson. 2004. "Can Language Restructure Cognition? The Case for Space." *Trends in Cognitive Science* 8:108–114.
Martinet, C., S. Valdois, & M. Fayol. 2004. "Lexical Orthographic Knowledge Develops from the Beginning of Literacy Acquisition." *Cognition* 91:B11–B22.
McLuhan, M. 1962. *The Gutenberg Galaxy.* Toronto: The University of Toronto Press.
——. 1964. *Understanding Media: The Extensions of Man.* New York: McGraw-Hill.

Medin, D. L., & S. Atran. 2004. "The Native Mind: Biological Categorization and Reasoning in Development across Cultures." *Psychological Review* 111:960–983.
Medin, D. L., E. B. Lynch, & K. O. Solomon. 2000. "Are There Kinds of Concepts?" *Annual Review of Psychology* 51:121–147.
Medin, D. L., B. H. Ross, & A. B. Markman. 2001. 3rd ed. *Cognitive Psychology*. New York: Harcourt.
Medin, D. L., N. O. Ross, S. Atran, D. Cox, J. Coley, J. B. Proffitt, & S. Blok. 2006. "Folkbiology of freshwater fish." *Cognition* 99:237–273.
Michalowski, P. 1994. "Writing and Literacy in Early States: A Mesopotamian Perspective." Pp. 49–70 in *Literacy: Interdisciplinary Conversations*, ed. D. Keller-Cohen. Cresskill, NJ: Hampton Press.
——. 1995. "Sumerian Literature: An Overview." Pp. 2279–2292 in *Civilizations of the Ancient Near East*, edited by J. Sasson. New York: Charles Scribner's Sons.
Miller, A. I. 1997. *Albert Einstein's Special Theory of Relativity: Emergence (1905) and Early Interpretation (1905–1911)*. Berlin: Springer.
Miller, G. A. 1956. "The Magical Number Seven Plus or Minus Two: Some Limits on Our Capacity for Processing Information." *Psychological Review* 63:81–97.
Miller, G. A., & P. Johnson-Laird. 1976. *Language and Perception*. Cambridge, MA: Harvard University Press.
Moorey, P. R. S. 1999. *Ancient Mesopotamian Materials and Industries: The Archaeological Evidence*. Winona Lake, IN: Eisenbrauns.
Münte, T. F., K. Schiltz, & M. Kutas. 1998. "When Temporal Terms Belie Conceptual Order." *Nature* 395:71–73.
Neugebauer, O. 1945. *Mathematical Cuneiform Texts*. New Haven: American Oriental Society.
——. 1955. *Astronomical Cuneiform Texts*. London: Lund Humphries, for the Institute of Advanced Studies, Princeton, NJ.
Neugebauer, O., & A. Sachs. 1967. "Some Atypical Astronomical Cuneiform Texts, I." *Journal of Cuneiform Studies* 21:183–217.
——. 1969. "Some Atypical Astronomical Cuneiform Texts, II." *Journal of Cuneiform Studies* 22:92–113.
Newton, I. 1686/7. *Philosophae Naturalis Principia Mathematica*. London: Joseph Slater.
Nisbett, R. 2003. *The Geography of Thought: How Asians and Westerners Think Differently, and Why*. New York: Free Press.
Nissen, H. J., P. Damerow, & R. K. Englund. 1993. *Archaic Bookkeeping: Early Writing and Techniques of Economic Administration in the Ancient Near East*. Translated by P. Larsen. Chicago: University of Chicago Press.
Nunberg, G. 1993. "Indexicality and Deixis." *Linguistics and Philosophy* 16:1–43.
Olson, D. R. 1977. "From Utterance to Text: The Bias of Language in Speech and Writing." *Harvard Educational Review* 47:257–281.
——. 1994. *The World on Paper*. Cambridge: Cambridge University Press.
——. 1996. "Towards a Psychology of Literacy: On the Relations between Speech and Writing." *Cognition* 60:83–104.
——. 2001. "What Writing Is." *Pragmatics and Cognition* 9:239–258.
Olson, D. R., & N. Torrance. 2009. *The Cambridge Handbook of Literacy*. Cambridge: Cambridge University Press.
Ong, W. 1982. *Orality and Literacy: The Technologizing of the Word*. London: Methuen.
Oppenheim, A. L. 1966. "Mesopotamia in the Early History of Alchemy." *Revue d'assyriologie* 60:29–45.
——. 1970. *Glass and Glassmaking in Ancient Mesopotamia*. Corning, NY: Corning Museum of Glass.
——. 1978. "Man and Nature in Mesopotamian Civilization." Pp. 634–666 in *Dictionary of Scientific Biography*, vol. 15. New York: Scribner.
Parpola, S. 1993. *Letters from Assyrian and Babylonian Scholars*. State Archives of Assyria 10. Helsinki: Helsinki University Press.

Pearce, L. E. 1995. "The Scribes and Scholars of Ancient Mesopotamia." Pp. 2265–2278 in *Civilizations of the Ancient Near East*, ed. J. Sasson. New York: Charles Scribner's Sons.

Peirce, C. S. 1873/1991. "On the Nature of Signs." Pp. 141–143 in *Peirce on Signs: Writings on Semiotic by Charles Sanders Peirce*, ed. J. Hoopes. Chapel Hill: University of North Carolina Press.

Perelman, C., & L. Olbrechts-Tyteca. 1969. *A New Rhetoric: A Treatise on Argumentation*. South Bend, IN: University of Notre Dame Press.

Pinker, S. 1994. *The Language Instinct*. New York: Morrow.

——. 2002. *The Blank Slate: The Modern Denial of Human Nature*. New York: Viking.

——. 2004. "Why Nature and Nurture Won't Go Away." *Daedalus* 133, no. 4:5–17.

Pinker, S., & R. Jackendoff. 2005. "The Faculty of Language: What's Special about It?" *Cognition* 95:201–236.

Pongratz-Leisten, B. 2010. "The Globalization of Knowledge in the Multi-Ethnic and Multi-Lingual Milieu of the Ancient Near East." Seminar and lecture series, Institute for the Study of the Ancient World, New York University.

Read, C. A., Y. Zhang, & B. Ding. 1986. "The Ability to Manipulate Speech Sounds Depends on Knowing Alphabetic Reading." *Cognition* 24:31–44.

Reiner, E. 1998. *Babylonian Planetary Omens, Part Three.* Groningen: Styx Publications.

Reiner, E., & D. Pingree. 1981. *Enūma Anu Enlil Tablets 50–51* (= *Babylonian Planetary Omens, Part Two* = *Bibliotheca Mesopotamica*, vol 2., fasc. 2). Malibu, CA: Undena Publications.

Renfrew, C. 1994. "Towards a Cognitive Archeology." Pp. 3–12 in *The Ancient Mind: Elements of Cognitive Archaeology*, edited by C. Renfrew & E. B. W. Zubrow. Cambridge: Cambridge University Press.

Reynolds, F. 1998. "Unpropitious Titles of Mars in Mesopotamian Scholarly Tradition." Pp. 347–358 in *Intellectual Life of the Ancient Near East: Papers Presented at the 43rd Rencontre assyriologique international, Prague, July 1–5, 1996*, edited by Jiři Proseck'y. Prague: Oriental Institute.

Richerson, P. J., & R. Boyd. 2005. *Not by Genes Alone: How Culture Transformed Human Evolution.* Chicago: University of Chicago Press.

Robins, R. H. 1997. *A Short History of Linguistics.* 4th ed. London: Longmans.

Robinson, R. 1954. *Definition.* Oxford: Clarendon Press.

Robson, E. 1998. *Mesopotamian Mathematics, 2100–1600 BC: Technical Constants in Bureaucracy and Education.* Oxford Editions of Cuneiform Texts 14. Oxford: Clarendon Press.

Rochberg, F. 2004. *The Heavenly Writing: Divination, Horoscopy, and Astronomy in Mesopotamian Culture.* Cambridge: Cambridge University Press.

——. 2008. "Babylonian Astronomy: The Hellenistic Transmission." *Mélanges de l'Université Saint-Joseph* 61:13–32.

——. 2009. "Conditionals, Inference, and Possibility in Ancient Mesopotamian Science." *Science in Context* 22:4–25.

——. In press. "If P, then Q: Form and Reasoning in Babylonian Divination." In *The Interpretation of Signs in Antiquity*, edited by A. Annus. Chicago: Oriental Institute Publications.

Rochberg-Halton, F. 1982. "Fate and Divination in Mesopotamia." Pp. 363–371 in *Vorträge gehalten auf der 28. Rencontre Assyriologique Internationale in Wien, 6.–10. Juli 1981*, edited by H. Hirsch & H. Hunger. Archiv für Orientforschung Beiheft 19. Horn, Austria: Ferdinand Berger.

Rubio, G. 2007. "Writing in Another Tongue: Alloglottography in the Ancient Near East." Pp. 33–70 in *Margins of Writing: Origins of Cultures*, edited by S. Sanders. Chicago: University of Chicago Press.

Schaffer, A. 1981. " 'Up' and 'down,' 'front' and 'back.' Gilgamesh, XI, 78 and Atrahasis III 20–31." *Revue d'assyriologie* 75:75–77.

Schmandt-Bessarat, D. 1986. "Tokens: Facts and Interpretations." *Visible Language* 20:250–272.

——. 1992. *Before Writing*. Austin: University of Texas Press.
Scribner, S., & M. Cole. 1981. *The Psychology of Literacy*. Cambridge, MA: Harvard University Press.
Shankweiler, D., & I. Liberman. 1972. "Misreading: A Search for Causes." Pp. 293–317 in *Language by Ear and by Eye: The Relationship between Speech and Reading*, edited by J. Kavanaugh & I. Mattingly. Cambridge, MA: MIT Press.
Siok, W. T., C. A. Perfetti, Z. Jin, & L. H. Tan. 2004. "Biological Abnormality of Impaired Reading is Constrained by Culture." *Nature* 431:71–76.
Soden, W. von. 1936. "Leistung und Grenze sumerischer und babylonischer Wissenschaft." *Die Welt als Geschichte* 2:411–462.
Soll, W. M. 1988. "Babylonian and Biblical Acrostics." *Biblica* 69:305–323.
Sommerfeld, W. 1995. "The Kassites of Ancient Mesopotamia: Origins, Politics, and Culture." Pp. 917–930 in *Civilizations of the Ancient Near East*, edited by J. Sasson. New York: Charles Scribner's Sons.
Spelke, E. S., & K. D. Kinzler. 2007. "Core Knowledge." *Developmental Science* 10: 89–96.
Spelke, E. S., & S. Tsivkin. 2001. "Initial Knowledge and Conceptual Change: Space and Number." Pp. 70–97 in *Language Acquisition and Conceptual Development*, edited by M. Bowerman & S. C. Levinson. Cambridge: Cambridge University Press.
Sperber, D. 1994. "Understanding Verbal Understanding." Pp. 179–198 in *What Is Intelligence?*, edited by J. Khalfa. Cambridge: Cambridge University Press.
——. 1996. *Explaining Culture: A Naturalistic Approach*. Oxford: Blackwell.
Sperber, D., & N. Claudiere. 2006. "Why Modeling Cultural Evolution Is Such a Challenge." *Biological Theory* 1:20–22.
Sperber, D., & L. A. Hirschfeld. 2004. "The Cognitive Foundations of Cultural Stability and Diversity." *Trends in Cognitive Science* 8:40–46.
Sperber, D., & D. Wilson. 1995. *Relevance: Communication and Cognition*. 2nd ed. Oxford: Blackwell.
——. 2002. "Pragmatics, Modularity, and Mind-Reading." *Mind & Language* 17: 3–23.
Stanovich, K. 2000. *Progress in Understanding Reading: Scientific Foundations and New Frontiers*. New York: Guilford Press.
Stock, B. 1983. *The Implications of Literacy*. Princeton, NJ: Princeton University Press.
Talmy, L. 1983. "How Language Structures Space." Pp. 225–282 in *Spatial Orientation: Theory, Research, and Application*, edited by H. Pick & L. Acredolo. New York: Plenum.
——. 2000. *Toward a Cognitive Semantics*. Cambridge, MA: MIT Press.
Tan, L. H., A. Chan, P. Kay, P. Khong, L. Yip, & K. Luke. 2008. "Language Affects Patterns of Brain Activation Associated with Perceptual Decision." *Proceedings of the National Academy of Science* 105:4004–4009.
Tan, L. H., A. R. Laird, K. Li, & P. T. Fox. 2005. "Neuroanatomical Correlates of Phonological Processing of Chinese Characters and Alphabetic Words: A Meta-Analysis." *Human Brain Mapping* 25:83–91.
Tigay J. 1982. *The Evolution of the Gilgamesh Epic*. Philadelphia: University of Pennsylvania Press.
Tomasello, M. 1999. *The Cultural Origins of Human Cognition*. Cambridge, MA: Harvard University Press.
——. 2003. *Constructing a Language*. Cambridge, MA: Harvard University Press.
Tomasello, M., M. Carpenter, & U. Liszkowski. 2007. "A New Look at Infant Pointing." *Child Development* 78:705–722.
Van Soldt, W. 1995. *Solar Omens of Enuma Anu Enlil: Tablets 23 (24)–29 (30)*. Istanbul: Nederlands Historisch-Archaeologisch Instituut.
Vanstiphout, H. 1995. "Memory and Literacy in Ancient Western Asia." Pp. 2181–2196 in *Civilizations of the Ancient Near East*, edited by J. Sasson. New York: Charles Scribner's Sons.

Veldhuis, N. 1997. "Elementary Education at Nippur: The Lists of Trees and Wooden Objects." PhD dissertation, University of Groningen.

——. 2004. *Religion, Literature, and Scholarship: The Sumerian Composition Nanše and the Birds, with a Catalogue of Sumerian Bird Names*. Cuneiform Monographs 22. Leiden: Brill/Styx.

Verderame, L. 2002. *Le Tavole I–VI della serie astrologica "Enūma Anu Enlil."* Nisaba 2. Rome: Di. Sc. A.M.

Walker, C. B. F. 1990. "Cuneiform." Pp. 15–73 in *Reading the Past: Ancient Writing from Cuneiform to the Alphabet*. London: The British Museum Press.

Watson, R. 1985. "Towards a Theory of Definition." *Journal of Child Language* 12: 181–197.

——. 1995. "Relevance and Definition." *Journal of Child Language* 22:211–222.

——. 2000. "Cognition and the Lexicon in the Environment of Texts." Pp. 62–79 in *Minds in the Making: Essays in Honor of David R. Olson*, edited by J. W. Astington. London: Blackwell.

Watson, R., & M. Blondheim, eds. 2007. *The Toronto School of Communication Theory: International Perspectives*. Toronto: University of Toronto Press; Jerusalem: The Hebrew University Magnes Press.

Watson, R., & D. R. Olson. 1987. "From Meaning to Definition: A Literate Bias on the Structure of Word Meaning." Pp. 329–353 in *Comprehending Oral and Written Language*, edited by R. Horowitz & J. Samuels. New York: Academic Press.

Waxman, S. R., & I. E. Braun. 2005. "Consistent (But Not Variable) Names as Invitations to Form Object Categories: New Evidence from 12-Month-Old Infants." *Cognition* 95:B59–B68.

Weidner, E. F. 1915. *Handbuch der babylonischen Astronomie*. Leipzig: J. C. Hinrich'sche Buchhandlung.

——. 1959–60. "Ein astrologischer Sammeltext aus der Sargonidenzeit." *Archiv für Orientforschung* 19:105–113.

Wellman, H. M., & S. A. Gelman. 1992. "Cognitive Development: Foundational Theories of Core Domains." *Annual Review of Psychology* 43:337–375.

Wittgenstein, L. 1955. *Philosophical Investigations*. Translated by G. E. M. Anscombe. Oxford: Basil Blackwell.

Zwaan, R. A. 1999. "Situation Models: The Mental Leap into Imagined Worlds." *Current Directions in Psychological Science* 8:15–18.

## APPENDIX ONE

# THE TRANSLATED TEXT OF MUL.APIN

From Hunger & Pingree, 1989
(¶ signifies the cuneiform marker DIŠ, which occurs in the transliteration but is omitted in the Hunger & Pingree translation)

Ii1 ¶ The Plow, Enlil, who goes at the front of the stars of Enlil.
Ii2 ¶ The Wolf, the seeder of the Plow.
Ii3 ¶ The Old Man, Enmešarra.
Ii4 The Crook, Gamlum.
Ii5 ¶ The Great Twins, Lugalgirra and Meslamtaea.
Ii6 ¶ The Little Twins, Alammuš and Nin-EZENxGUD (Gublaga).
Ii7 ¶ The Crab, the seat of Anu.
Ii8 ¶ The Lion, Latarak.
Ii9 ¶ The star which stands in the breast of the Lion: the King.
Ii10 ¶ The dusky stars which stand in the tail of the Lion:
Ii11 The Frond (of the date palm) of Eru, Zarpanitu.
Ii12 ¶ ŠU.PA, Enlil who decrees the fate of the land.
Ii13 ¶ The star which stands in front of it: the Abundant One, the messenger of Ninlil.
Ii14 ¶ The star which stands behind it: the Star of Dignity, the messenger of Tišpak.
Ii15 ¶ The Wagon, Ninlil.
Ii16 ¶ The star which stands in the cart-pole of the Wagon:
Ii17 The Fox, Erra, the strong one among the gods.
Ii18 ¶ The star which stands in front of the Wagon: the Ewe, Aya.
Ii19 ¶ The Hitched Yoke, the great Anu of Heaven.
Ii20 ¶ The Wagon of Heaven, Damkianna.
Ii21 ¶ The star which stands in its rope: the Heir of the Sublime Temple,
Ii22 the first-ranking son of Anu.
Ii23 ¶ The Standing Gods of Ekur, the Sitting Gods of Ekur.
Ii24 ¶ The She-Goat, Gula.
Ii25 ¶ The star which stands in front of the She-Goat: the Dog.

Ii26 ¶ The bright star of the She-Goat: Lamma, the messenger of Baba.
Ii27 ¶ The two stars which stand behind it: Nin-SAR and Erragal.
Ii28 ¶ The Panther: Nergal.
Ii29 ¶ The star which stands at its right side: the Pig, Damu.
Ii30 ¶ The star which stands at its left side: the Horse.
Ii31 ¶ The star which stands behind it: the Stag, the messenger of the Stars.
Ii32 ¶ The dusky stars which stand in the breast of the Stag:
Ii33 Harriru, the Rainbow.
Ii34 ¶ The bright red star which stands in the kidney of the Stag:
Ii35 The Deleter.
Ii36 When the stars of Enlil have been finished,
Ii37 one big star—(although) its light is dim—divides the sky in half and stands there: (that is) the star of Marduk, the Ford,
Ii38 ¶ Jupiter, (it) keeps changing its position and crosses the sky.

---

Ii39 33 stars of Enlil

---

Ii40 ¶ The Field, the seat of Ea, which goes at the front of the stars of Anu.
Ii41 ¶ The star which stands opposite the Field: the Swallow.
Ii42 ¶ The star which stands behind the Field, Anunitu.
Ii43 ¶ The star which stands behind it: the Hired Man, Dumuzi.
Ii44 ¶ The Stars, the seven gods, the great gods.
Iii1 ¶ The Bull of Heaven, the Jaw of the Bull, the crown of Anu.
Iii2 ¶ The True Shepherd of Anu, Papsukal, the messenger of Anu and Ištar.
Iii3 ¶ The twin stars which stand opposite the True Shepherd of Anu:
Iii4 Lulal and Latarak.
Iii5 ¶ The star which stands behind it: the Rooster.
Iii6 ¶ The Arrow, the arrow of the great warrior Ninurta.
Iii7 ¶ The Bow, the Elamite Ištar, the daughter of Enlil.
Iii8 ¶ The Snake, Ningizzida, lord of the Netherworld.
Iii9 ¶ The Raven, the star of Adad.
Iii10 ¶ The Furrow, Šala, the ear of corn.
Iii11 ¶ The Scales, the horn of the Scorpion.
Iii12 ¶ The star of Zababa, the Eagle, and the Dead Man.
Iii13 ¶ Venus keeps changing its position and crosses the sky.

Iii14 ¶ Mars keeps changing its position and crosses the sky.

Iii15 ¶ Saturn keeps changing its position and crosses the sky.

Iii16 ¶ Mercury, whose name is Ninurta, rises or sets in the east

Iii17 or in the west within a month.

---

Iii18 23 stars of Anu.

---

Iii19 ¶ The Fish, Ea, who goes at the front of the stars of Ea.

Iii20 ¶ The Great One, Ea; the star of Eridu, Ea.

Iii21 ¶ The star which stands at its right: Ninmah.

Iii22 ¶ EN.TE.NA.BAR.HUM, Ningirsu.

Iii23 ¶ The star which stands at its side: The Harrow, the weapon of Mar-biti,

Iii24 inside of which one sees the subterranean waters.

Iii25 ¶ The two stars which stand behind it: Šullat and Haniš, Šamaš and Adad.

Iii26 ¶ The star which stands behind them rises like Ea

Iii27 and sets like Ea: Numušda, Adad.

Iii28 ¶ The star which stands at the left side of the Scorpion: the Mad Dog, Kusu.

Iii29 ¶ The Scorpion, Išhara, goddess of all inhabited regions.

Iii30 ¶ The Breast of the Scorpion: Lisi, Nabû.

Iii31 ¶ The two stars which stand in the sting of the Scorpion:

Iii32 Šarur and Šargaz.

Iii33 ¶ The star which stands behind them: Pabilsag.

Iii34 ¶ The Bark and the Goat-Fish.

---

Iii35 15 stars of Ea.

---

Iii36 ¶ On the 1st of Nisannu the Hired Man becomes visible.

Iii37 ¶ On the 20th of Nisannu the Crook becomes visible.

Iii38 ¶ On the 1st of Ajjaru the Stars become visible.

Iii39 ¶ On the 20th of Ajjaru the Jaw of the Bull becomes visible.

Iii40 ¶ On the 10th of Simanu the True Shepherd of Anu and the Great Twins become visible.

Iii41 ¶ On the 5th of Du'uzu the Little Twins and the Crab become visible.

Iii42 ¶ On the 15th of Du'uzu the Arrow, the Snake, and the Lion

Iii43 become visible; 4 minas is a daytime watch, 2 minas is a nighttime watch.

Iii44 ¶ On the 5th of Abu the Bow and the King become visible.
Iii44a ¶ On the 1st of Ululu [....][1]
Iii45 ¶ On the 10th of Ululu the star of Eridu and the Raven become visible.
Iii46 ¶ On the 15th of Ululu ŠU.PA, Enlil, becomes visible.
Iii47 ¶ On the 25th of Ululu the Furrow becomes visible.
Iiii1 ¶ On the 15th of Tešritu the Scales, the Mad Dog, EN.TE. NA.BAR.HUM,
Iiii2 and the Dog become visible; 3 minas is a daytime watch, 3 minas is a nighttime watch.
Iiii3 ¶ On the 5th of Arahsamnu the Scorpion becomes visible.
Iiii4 ¶ On the 15th of Arahsamnu the She-Goat and the Breast of the Scorpion become visible.
Iiii5 ¶ On the 15th of Kislimu the Panther, the Eagle,
Iiii6 and Pabilsag become visible.
Iiii7 ¶ On the 15th of Ṭebetu SIM.MAH, (i.e.) the Swallow (or) IM.ŠEŠ,
Iiii8 becomes visible in the East, and the Arrow
Iiii9 becomes visible in the evening; 2 minas is a daytime watch, 4 minas is a nighttime watch.
Iiii10 ¶ On the 5th of Šabaṭu the Great One, the Field, and the Stag become visible.
Iiii11 ¶ On the 25th of Šabaṭu Anunitu becomes visible.
Iiii12 ¶ On the 15th of Addaru the Fish and the Old Man become visible.

---

Iiii13 ¶ The Stars rise and the Scorpion sets.
Iiii14 ¶ The Scorpion rises and the Stars set.
Iiii15 ¶ The Bull of Heaven rises and ŠU.PA sets.
Iiii16 ¶ The True Shepherd of Anu rises and Pabilsag sets.
Iiii17 ¶ The Arrow, the Snake, and the Lion rise, and
Iiii18 the Great One and the Eagle set.
Iiii19 ¶ The Bow and the King rise, and the She-Goat sets.
Iiii20 ¶ The star of Eridu and the Raven rise, and the Panther sets.

[1] This appears as an extra line in only two manuscripts and is not numbered in the Hunger-Pingree edition.

Iiii21 ¶ ŠU.PA, Enlil, rises and the Field sets.
Iiii22 ¶ Ninmah rises and Anunitu sets.
Iiii23 ¶ The Scales, the Mad Dog, and EN.TE.NA.BAR.HUM
Iiii24 rise, and the Hired Man sets.
Iiii25 ¶ The Scorpion and the Dog rise, and the Star of Eridu and the Stars set.
Iiii26 ¶ The Breast of the Scorpion and the She-Goat rise, and the Old Man and the True Shepherd of Anu set.
Iiii27 ¶ Pabilsag, Zababa, and the Standing Gods rise, and
Iiii28 the Arrow, the Bow, and the Crook set.
Iiii29 ¶ The Panther and the Eagle rise, and
Iiii30 the Great Twins and the Little Twins set.
Iiii31 ¶ The Field, the Great One, and the Stag rise and
Iiii32 the Lion, the Snake, and EN.TE.NA.BAR.HUM set.
Iiii33 ¶ The Fish and the Old Man rise, and the Furrow and the Mad Dog set.

---

Iiii34 ¶ 55 days pass from the rising of the Arrow to the rising of the star of Eridu.
Iiii35 ¶ 60 days pass from the rising of the Arrow to the rising of ŠU.PA.
Iiii36 ¶ 10 days pass from the rising of ŠU.PA to the rising of the Furrow.
Iiii37 ¶ 20 days pass from the rising of the Furrow to the rising of the Scales.
Iiii38 ¶ 30 days from the rising of the Scales to the rising of the She-goat.
Iiii39 ¶ 30 days pass from the rising of the She-Goat to the rising of the Panther.
Iiii40 ¶ 30 days pass from the rising of the Panther to the rising of the Swallow.
Iiii41 ¶ 20 days pass from the rising of the Swallow to the rising of the Field.
Iiii42 ¶ 40 days pass from the rising of the Field to the rising of the Fish.
Iiii43 ¶ 35 days pass from the rising of the Fish to the rising of the Crook.
Iiii44 ¶ 10 days pass from the rising of the Crook to the rising of the Stars.

Iiii45 ¶ 20 days pass from the rising of the Stars to the rising of the Bull of Heaven.

Iiii46 ¶ 20 days pass from the rising of the Bull of Heaven to the rising of the True Shepherd of Anu.

Iiii47 ¶ 35 days pass from the rising of the True Shepherd of Anu to the rising of the Arrow.

Iiii48 ¶ 20 days pass from the rising of the Arrow to the rising of the Bow.

---

Iiii49 The stars enter into the night in the morning 1 UŠ each day.

Iiii50 The stars come out into the day in the evening 1 UŠ each day.

---

Iiv1 ¶ The *ziqpu* stars which stand in the path of Enlil in the middle of the sky

Iiv2 opposite the breast of the observer of the sky, and

Iiv3 by means of which he observes the rising and setting of the stars at night (are the following):

---

Iiv4 ¶ ŠU.PA, the star of Dignity, the Standing Gods, the Dog,

Iiv5 the She-Goat, the Panther, the Stag, the Old Man, the Crook,

Iiv6 the Great Twins, the Crab, the Lion, Eru, and the Abundant One.

---

Iiv7 All these are the *ziqpu* stars in the path of the stars of Enlil which stand in the middle of the sky

Iiv8 opposite your breast, and by means of which you observe

Iiv9 the risings and settings of the stars at night.

---

Iiv10 If you are to observe the *ziqpu*, you stand

Iiv11 in the morning before sunrise, West to your right,

Iiv12 East to your left, your face directed towards South;

Iiv13 on the 20th of Nisannu the *kumāru* of the Panther stands in the middle of the sky

Iiv14 opposite your breast, and the Crook rises.

---

Iiv15 ¶ On the 1st of Ajjaru, the Breast of the Panther stands in the middle of the sky

Iiv16 opposite your breast, and the stars rise.

---

Iiv17 ¶ On the 20th of Ajjaru, the Knee of the Panther
Iiv18 stands in the middle of the sky opposite your breast, and the Jaw of the Bull rises.

---

Iiv19 ¶ On the 10th of Simanu the Heel of the Panther stands in the middle of the sky
Iiv20 opposite your breast, and the True Shepherd of Anu rises.

---

Iiv21 ¶ On the 15th of Du'uzu the bright star of the Old Man stands in the middle of the sky opposite your breast, and the Arrow rises.

---

Iiv22 ¶ On the 15th of Abu the dusky stars of the Old Man
Iiv23 stand in the middle of the sky opposite your breast, and the Bow rises.

---

Iiv24 ¶ On the 15th of Ululu the Great Twins stand in the middle of the sky opposite your breast, and ŠU.PA and the Star of Eridu rise.

---

Iiv25 ¶ On the 15th of Tešritu the Lion stands in the middle of the sky opposite your breast, and the Scales rise.

---

Iiv26 ¶ On the 15th of Arahsamnu Eru stands in the middle of the sky opposite your breast, and the She-Goat rises.

---

Iiv27 ¶ On the 15th of Kislimu ŠU.PA stands in the middle of the sky opposite your breast, and the Panther rises.

---

Iiv28 ¶ On the 15th of Ţebetu the Standing Gods stand in the middle of the sky opposite your breast, and the Swallow rises.

---

Iiv29 ¶ On the 15th of Šabaţu the Dog stands in the middle of the sky opposite your breast, and the Field rises.

---

Iiv30 ¶ On the 15th of Addaru the She-Goat stands in the middle of the sky opposite your breast, and the Fish rises.

---

Iiv31 The gods who stand in the path of the Moon, through whose regions the Moon in the course of a month
Iiv32 passes and whom he touches:

---

Iiv33 The Stars, the Bull of Heaven, the True Shepherd of Anu, the Old Man,
Iiv34 the Crook, the Great Twins, the Crab, the Lion,
Iiv35 the Furrow, the Scales, the Scorpion, Pabilsag,
Iiv36 the Goat-Fish, the Great One, the Tails, the Swallow,
Iiv37 Anunitu, and the Hired Man.

---

Iiv38 All these are the gods who stand in the path of the Moon, through whose regions the Moon in the course of a month
Iiv39 passes and whom he touches.

---

IIi1 ¶ The Sun travels the (same) path the Moon travels.
IIi2 ¶ Jupiter travels the (same) path the Moon travels.
IIi3 ¶ Venus travels the (same) path the Moon travels.
IIi4 ¶ Mars travels the (same) path the Moon travels.
IIi5 ¶ Mercury whose name is Ninurta travels the (same) path the moon travels.
IIi6 ¶ Saturn travels the (same) path the Moon travels.

---

IIi7 Together six gods who have the same position, (and) who touch the stars of the sky and
IIi8 keep changing their positions.

---

IIi9 ¶ On the 15th of Du'uzu the Arrow becomes visible, and
IIi10 4 minas is a daytime watch, 2 minas is a nighttime watch.
IIi11 The Sun which rose towards the North with the head of the Lion
IIi12 turns and keeps moving down towards the South at a rate of 40 NINDA per day.
IIi13 The days become shorter, the nights longer.

---

IIi14 ¶ On the 15th of Tešrītu the Sun rises in the Scales in the East,
IIi15 and the Moon stands in front of the Stars behind the Hired Man, 3 minas is a daytime watch, 3 minas is a nighttime watch.

---

IIi16 ¶ On the 15th of Ṭebetu, the Arrow becomes visible in the evening. 2 minas is a daytime watch, 4 minas is a nighttime watch.

IIi17 The Sun which rose towards the South with the head of the Lion turns and
IIi18 keeps coming up towards the North at a rate of 40 NINDA per day. The days become longer, the nights become shorter.

---

IIi19 ¶ On the 15th of Nisannu the Moon stands in the evening in
IIi20 the Scales in the East, and the Sun in the West in front of the Stars behind
IIi21 the Hired Man. 3 minas is a daytime watch. 3 minas is a nighttime watch.

---

IIi22 ¶ On the 15th of Nisannu, on the 15th of Du'uzu, on the 15th of Tešritu, on the 15th of Ţebetu,
IIi23 you observe the risings of the Sun, the visibility time of the Moon, the appearances of the Arrow,
IIi24 and you will find how many days are in excess.

---

IIi25 ¶ On the 10th of Ululu the star of Eridu becomes visible, on the 15th ŠU.PA;
IIi26 on the day their stars become visible you observe their risings, their glow, and
IIi27 their…., and the wind that blows; you guard (?) the horses
IIi28 so that they do not drink water from the river.
IIi29 When their stars have been made visible,
IIi30 you present offerings to them; horses
IIi31 will touch bitumen and drink water from the river.

---

IIi32 ¶ On the 5th of Arahsamnu the Scorpion becomes visible, on the 15th the breast of the Scorpion;
IIi33 on the day they become visible you observe the wind that blows.

---

IIi34 ¶ On the 15th of Addaru the Fish [becomes visible] in the morning, in the evening the star of Eridu [becomes visible….]
IIi35 their stars…. [….];
IIi36 on the day they become visible [you observe] their risings, their glow,
IIi37 their…., and the wind that blows.

---

IIi38 Jupiter, Venus, Mercury, whose name is Ninurta, Mars, Saturn,
IIi39 [also called] "the Scales" (or) "Star of the Sun."

---

IIi40 [These are the gods (?) who] keep changing their positions and their glow
IIi41 [and] touch [the stars of the sky]; on the day their stars become visible, you observe their risings, their glow,
IIi42 [their....], where they become visible, and the wind that blows: on the day they become visible,
IIi43 you present offerings to them; horses will touch bitumen.

---

IIi44 ¶ Venus disappears in the East and remains (invisible) in the sky for a month,
IIi45 or for a month and 15 days, or it remains for 2 months, and
IIi46 becomes visible in the West.

---

IIi47 ¶ Venus disappears in the West and becomes visible (again) in the East on the day she disappears; or, for 3 days,
IIi48 thirdly, for seven days, fourthly, for 14 days, she remains (invisible) and then rises.

---

IIi49 ¶ Jupiter disappears in the West and remains (invisible) in the sky for 20 days,
IIi50 or remains for a month, and rises and becomes visible in the East in the path of the Sun.

---

IIi51 ¶ Mars disappears in the West and remains (invisible) in the sky for 2 months, or for 3 months and 10 days, or for 6 months
IIi52 and 20 days, and rises and becomes visible in the East in the path of the Sun.

---

IIi53 ¶ Saturn disappears in the West, remains (invisible) in the sky for 20 days, and becomes visible in the path of the Sun.

---

IIi54 ¶ Mercury, whose name is Ninurta, becomes visible either in the East or in the West, and

IIi55 stands in the sky for 7 days, or for 14 days, thirdly, for 21 days, fourthly, for a month, fifthly, for a month and 15 days,
IIi56 and when it disappears it remains (invisible) for as many days as it stood in the sky
IIi57 and rises and becomes visible either in the East or in the West in the path of the Sun. (If) this star becomes visible in winter, (there will be) rain and flood;
IIi58 (if) it becomes visible at harvest time, you observe its glow, its…., where it becomes visible, and the wind that blows.
IIi59 This star is either red and bright, or yellow and dark.

---

IIi60 ¶ Jupiter becomes visible in the East, stands in the sky for one year, and disappears in the West.

---

IIi61 ¶ Venus becomes visible either in the East or in the West, stands in the sky for 9 months, and disappears.

---

IIi62 ¶ Mars becomes visible in the East, stands in the sky for one year and 6 months, or for one year and 10 months, or for 2 years,
IIi63 and disappears in the West. This star shows either redness and is bright, or is….and small.

---

IIi64 ¶ Saturn, also called the Scales (or) star of the Sun, becomes visible in the East,
IIi65 stands in the sky for one year, and disappears in the West. This star is either red or white.

---

IIi66 ¶ Mercury, whose name is Ninurta, becomes visible and disappears
IIi67 within a month either in the East or in the West.

---

IIi68 ¶ If you are to observe the direction of the winds: the Wagon lies across where the North wind rises,
IIi69 the Fish lies across where the South wind rises, the Scorpion lies across where the West wind rises,
IIi70 the Old Man and the Stars stand where the East wind rises.
IIi71 On the day of your observation the stars will indicate to you which wind blows.

---

IIGapA1 ¶ From the 1st of Addaru until the 30th of Ajjaru the Sun stands in the path of the Anu stars;
IIGapA2 wind and weather.

---

IIGapA3 ¶ From the 1st of Simanu until the 30th of Abu the Sun
IIGapA4 stands in the path of the Enlil stars; harvest and heat.

---

IIGapA5 ¶ From the 1st of Ululu until the 30th of Arahsamnu the Sun
IIGapA6 stands in the path of the Anu stars; wind and weather.

---

IIGapA7 ¶ From the 1st of Kislimu until the 30th of Šabaṭu the Sun stands in the path of the Ea stars; cold.

---

IIGapA8 [¶ If on the 1st of Nissanu] the Stars and the Moon are in conjunction, this year is normal.
IIGapA9 [¶ If] on the 3rd [of Nissanu] the Stars and the Moon are in conjunction, this year is a leap year.

---

IIGapA10 [¶ If] the Stars become visible [on the 1st of Ajjaru], this year is normal.
IIGapA11 [¶ If] the Stars become visible on the 1st of [Simanu], this year is a leap year.

---

IIGapA12 [¶ If on the 15th of Du'uzu] .... [....]
IIGapA13 [¶ If] the Arrow becomes visible [on the 15th of Abu], this year is [a leap year.]

---

IIGapA14 [¶ If] ŠU.PA becomes visible [on the 15th of Ululu], this year is [normal.]
IIGapA15 [¶ If] ŠU.PA becomes visible [on the 15th of Tešritu], this year is [a leap year.]

---

IIii1 [¶ If] the Stars and the Moon are in conjunction [on the 15th of Arahsamnu], this year is normal.
IIii2 [¶ If] the Stars and the Moon are in conjunction on the 15th of [Kislimu], this year is a leap year.

---

IIii3 [¶ If] the Arrow [becomes visible] in the East in the evening on the 15th of Ṭebetu, [when (?) the Sun] rises towards the South, turns and keeps coming up towards the North, this year is normal.

IIii4 [¶ If] the Arrow becomes visible in the evening on the 15th of Šabaṭu, this year is a leap year.

---

IIii5 [¶ If] the Fish and the Old Man become visible on the 15th of Addaru, this year is normal.

IIii6 [¶ If] the Fish and the Old Man become visible on the 15th of Nisannu, this year is a leap year.

---

IIii7 You lo[ok(?)] for the rising (?) and.... of the stars of Ea, Anu, and Enlil

IIii8 and name this year;

IIii9 when...., you compute [....] and year, and

IIii10 for the third year you make a prediction, and proclaim this year a leap year.

---

IIii11 ¶ To.... the day of disappearance of the Moon for 12 months, you proclaim an intercalary month in three years (variant: the third year);

IIii12 10 additional days in 12 months is the amount for one year.

---

IIii13 If you are to find the correction for day, month, and year:

IIii14 you multiply 1, 40, the correction for a day, by one month,

IIii15 and you find 50, the correction for one month; you multiply 50, the correction for one month,

IIii16 by 12 months, and you find 10 additional days, the amount for one year.

IIii17 In three years (variant: the third year) you proclaim (this year) a leap year.

---

IIii18 ¶ An intercalary Nisannu (belongs to) the reign of Šulgi;

IIii19 ¶ an intercalary Addaru (belongs to) the reign of Amurru;

IIii20 ¶ an intercalary Ululu (belongs to) the reign of the Kassites.

---

IIii21 ¶ On the 15th of Nisannu, 3 minas is a daytime watch, 3 minas is a nighttime watch.

| | | | |
|---|---|---|---|
| IIii22 | | 1 cubit of shadow | 2½ bēru of daytime |
| IIii23 | | 2 cubits of shadow | 1 bēru 7 UŠ 30 NINDA daytime |
| IIii24 | | 3 cubits of shadow | ½ bēru 5 UŠ daytime |
| IIii25 | ¶ | On the 15th of Du'uzu, 4 minas is a daytime watch, 2 minas is a nighttime watch. | |
| IIii26 | | 1 cubit of shadow | 2 bēru daytime |
| | | 2 cubits of shadow | [1] bēru daytime |
| IIii27 | | 3 cubits of shadow | ⅔ bēru daytime |
| | | 4 cubits of shadow | ½ bēru daytime |
| IIii28 | | 5 cubits of shadow | 12 UŠ daytime |
| | | 6 cubits of shadow | 10 UŠ daytime |
| IIii29 | | 8 cubits of shadow | 7 UŠ 30 NINDA daytime |
| IIii30 | | 9 cubits of shadow | 6 UŠ 40 NINDA daytime |
| | | 10 cubits of shadow | 6 UŠ daytime |
| IIii31 | ¶ | On the 15th of Tešritu, 3 minas is a daytime watch, 3 minas is a nighttime watch. | |
| IIii32 | | 1 cubit of shadow | 2½ bēru daytime |
| IIii33 | | 2 cubits of shadow | 1 bēru 7 UŠ 30 NINDA daytime |
| IIii34 | | 3 cubits of shadow | ⅔ bēru 5 UŠ daytime |
| IIii35 | ¶ | On the 15th of Ţebetu, 2 minas is a daytime watch, 4 minas is a nighttime watch. | |
| IIii36 | | 1 cubit of shadow | 3 bēru daytime |
| | | 2 cubits of shadow | 1½ bēru daytime |
| IIii37 | | 3 cubits of shadow | 1 bēru daytime |
| | | 4 cubits of shadow | ⅔ bēru 2 UŠ 30 NINDA daytime |
| IIii38 | | 5 cubits of shadow | 18 UŠ daytime |
| | | 6 cubits of shadow | ½ bēru daytime |
| IIii39 | | 8 cubits of shadow | 11 UŠ 15 NINDA daytime |
| | | 9 cubits of shadow | 10 UŠ daytime |
| IIii40 | | 10 cubits of shadow | 9 UŠ daytime |
| IIii41 | | If you are to find the difference for 1 cubit of shadow, you multiply 40, the difference for daytime | |
| IIii42 | | and nighttime, by 7, 30, and you find 5, the difference for 1 cubit of shadow. | |

IIii43 ¶ On the 1st of Nisannu a nighttime watch is 3 minas 10 shekels; 12 UŠ 40 NINDA setting of the Moon.
IIii44 ¶ On the 15th of Nisannu a nighttime watch is 3 minas; 12 UŠ rising of the Moon.

---

IIii45 ¶ On the 1st of Ajjaru a nighttime watch is $2^5/_6$ minas; 11 UŠ 20 NINDA setting of the Moon.
IIii46 ¶ On the 15th of Ajjaru a nighttime watch is $2^2/_3$ minas; 10 UŠ 40 NINDA rising of the Moon.

---

IIii47 ¶ On the 1st of Simanu a nighttime watch is $2^1/_2$ minas; 10 UŠ setting of the Moon.
IIii48 ¶ On the 15th of Simanu a nighttime watch is $2^1/_3$ minas; 9 UŠ 20 NINDA rising of the Moon.

---

IIii49 ¶ On the 1st of Du'uzu a nightime watch is 2 minas 10 shekels; 8 UŠ 40 NINDA setting of the Moon.
IIii50 ¶ On the 15th of Du'uzu a nighttime watch is 2 minas; 8 UŠ [rising of the Moon].

---

IIii51 ¶ On the 1st of Abu a nighttime watch is 2 minas 10 shekels; [8 UŠ 40 NINDA setting of the Moon].
IIii52 ¶ On the 15th of Abu a nighttime watch is $2^1/_3$ minas; 9 U[Š 20 NINDA rising of the Moon].

---

IIii53 ¶ On the 1st of Ululu a nighttime watch is $2^1/_2$ minas; 10 UŠ setting [of the Moon].
IIii54 ¶ On the 15th of Ululu a nighttime watch is $2^2/_3$ minas; 10 UŠ 40 NINDA rising [of the Moon].

---

IIiii1 ¶ On the 1st of Tešritu a nighttime watch is $2^5/_6$ minas; 11 UŠ 20 NINDA setting [of the Moon].
IIiii2 ¶ On the 15th of Tešritu a nighttime watch is 3 minas; 12 UŠ rising [of the Moon].

---

IIiii3 ¶ On the 1st of Arahsamnu a nightime watch is 3 minas 10 shekels; 12 UŠ 40 NINDA setting of [the Moon].
IIiii4 ¶ On the 15th of Arahsamnu a nighttime watch is $3^1/_3$ minas; 13 UŠ 20 NINDA rising of the Moon.

---

IIiii5 ¶ On the 1st of Kislimu a nighttime watch is $3\frac{1}{2}$ minas; 14 UŠ setting of the Moon.

IIiii6 ¶ On the 15th of Kislimu a nightime watch is $3\frac{2}{3}$ minas; 14 UŠ 40 NINDA rising of the Moon.

---

IIiii7 ¶ On the 1st of Ţebetu a nighttime watch is $3\frac{5}{6}$ minas; 15 UŠ 20 NINDA setting of the Moon.

IIiii8 ¶ On the 15th of Ţebetu a nighttime watch is 4 minas; 16 UŠ rising of the Moon.

---

IIiii9 ¶ On the 1st of Šabaţu a nighttime watch is $3\frac{5}{6}$ minas; 15 UŠ 20 NINDA setting of the Moon.

IIiii10 ¶ On the 15th of Šabaţu a nighttime watch is $3\frac{1}{2}$ minas; 14 UŠ 40 NINDA rising of the Moon.

---

IIiii11 ¶ On the 1st of Addaru a nighttime watch is $3\frac{1}{2}$ minas; 14 UŠ setting of the Moon.

IIiii12 ¶ On the 15th of Addaru a nighttime watch is $3\frac{1}{3}$ minas; 13 UŠ 20 NINDA rising of the Moon.

---

IIiii13 4 is the coefficient for the visibility of the Moon; you multiply 3 minas, a nighttime watch,

IIiii14 by 4, and you find 12, the visibility of the Moon.

IIiii15 You multiply 40 NINDA, the difference for daytime and nighttime, by 4, and you find 2, 40, the difference of the visibility.

---

IIiii16 ¶ If the light (?) of the Stars is red (?): the irrigated land will prosper.

IIiii17 ¶ If the constellation of Ištar....s in Addaru: ....

IIiii18 ¶ If one star in the constellation of Ištar is very bright: the enemy.

IIiii19 ¶ If....4 or 2 big ones are yellow: deaths (?).

IIiii20 ¶ If all....are very red: flood and rain.

IIiii21 ¶ If the Fish in [....] either shows redness and is bright, or is....and small: ....

---

IIiii22 ¶ If the U.RI.RI-star becomes visible: rain and flood.

IIiii23 ¶ If the U.RI.RI-star approaches the Chariot: horses will die.

IIiii24 ¶ If the U.RI.RI-star approaches the Furrow: the same.
IIiii25 ¶ If the U.RI.RI-star approaches the Crab: ....will die.
IIiii26 ¶ If the U.RI.RI-star approaches the Crab on the right side: ....will die.
IIiii27 ¶ If the U.RI.RI-star approaches the Crab on the left side: ....will die.

---

IIiii28 ¶ If the Raven....s below to the direction of the South wind: sesame will prosper.
IIiii29 ¶ If the Raven...s above to the direction of the North wind: the barley crop will not prosper.

---

IIiii30 ¶ If the stars of the Lion.... : the king will be victorious wherever he goes. This star
IIiii31 is a campaign star.
IIiii32 ¶ If the KAL.NE-star....s and approaches the 4 stars of the Stars: the ruler (?) of the city will become good (?).
IIiii33 ¶ If Jupiter is bright: rain and flood.
¶ The name of the Rainbow is "day (?) of abundance";
IIiii34 ¶ In the South, rain; in the North, flood; in the East, rain; in the West, devastation.

---

IIiii35 ¶ On the day the Lisi-star becomes visible, a man should wake up at night all that is around his house,
IIiii36 people, cattle, sheep, donkeys,
IIiii37 and he must not sleep; he should pray to the Lisi-god, then he and all that is around his house will experience success.

---

IIiii38 ¶ If in Kislimu, Ṭebetu, or Šabaṭu the left horn of the Moon is pointed
IIiii39 and looks towards the earth: .....

---

IIiii40 ¶ If the Sun rises in a *nīdu*-cloud: the king will become furious and raise weapons.
IIiii41 ¶ If the sun sets in a *nīdu*-cloud: the king will die.

---

IIiii42 ¶ If a star flares up from the West and enters in the Lisi-star: there will be revolution.
IIiii43 ¶ If a star flares up from the West and enters the Yoke: there will be revolution.

IIiii44 ¶ If a star flares up from the West and enters the Moon: there will be revolution;

IIiii45 [¶ If] this star comes out (from the Moon) as three stars: unsuccessful attack.

---

IIiii46 If a star flashes from the East towards the South, passes EN.TE.NA.BAR.HUM

IIiii47 and sets in the West: for three years the land will see abundance.

IIiii48 [¶ If a star] passes from the West to the East: for three years

IIiii49 the land will experience evil.

IIiii50 ¶ If a star flares up from the middle of the sky and sets in the West: a heavy loss

IIiii51 will occur in the land.

IIiii52 [¶ If.....] the Scorpion becomes visible and the South wind blows: this year will be good.

---

IIiii53 [¶ If.....] ....: the king of the West will prosper.

---

Gap B 1 [¶ If] the star [of Mar]duk becomes visible at the beginning of the year: in this year the crop will prosper.

Gap B 2 [¶ If] the star of Marduk reaches the Stars: in this year the Storm god will devastate.

Gap B 3 ¶ If the star of Marduk reaches the Raven: early sesame will prosper.

Gap B 4 ¶ If the star of Marduk sees the body of a man: epilepsy (?) will seize him.

Gap B 5 ¶ If a man takes a bath in front of the star of Marduk: there will be guilt.

Gap B 6 ¶ If the star of Marduk is dark when it becomes visible: in this year there will be *asakku*-disease.

---

Gap B 7 ¶ If the Yoke is dim when it comes out: the late flood will come.

Gap B 8 ¶ If the Yoke keeps flaring up when it comes out: the flood will be early.

IIiv1 ¶ If the Yoke keeps flaring up like fire when it comes out: the crop will prosper.

IIiv2 ¶ If the Yoke is very low and dim when it comes out: there will be no flood.

IIiv3 ¶ If the Yoke is turned towards sunset when it comes out, (if) the West wind blows and
IIiv4 turns to South: on the 10th of Ululu there will be destruction of the land.
IIiv5 ¶ If the Yoke is turned towards sunrise (?) when it comes out and faces the front of the sky,
IIiv6 and no wind blows: there will be famine, the dynasty will disappear;
IIiv7 omen of Ibbi-Sin, king of Ur, who went in fetters
IIiv8 to Anšan; after him his people weep (variant: fall).

---

IIiv9 ¶ If a man is made a ruler, and the South wind blows: this man will become good.
IIiv10 ¶ If a man is made a ruler, and the North wind blows: he will eat thin (?) bread.
IIiv11 ¶ If a man is made a ruler, and the East wind blows: his days will be short.
IIiv12 ¶ If a man is made a ruler, and the West wind blows: he will not prosper.

APPENDIX TWO

# THE BABYLONIAN MONTH-NAMES

| Babylonian name* | Hebrew equivalent | Modern equivalent |
|---|---|---|
| *Nisannu* | Nisan | March/April |
| *Ajjaru* | Iyar | April/May |
| *Simānu* | Sivan | May/June |
| *Du'ūzu* | Tammuz | June/July |
| *Abu* | Ab | July/August |
| *Ulūlu* | Elul | August/September |
| *Tešrītu* | Tishrei | September/October |
| *Arahsamnu* | Marcheshvan | October/November |
| *Kislīmu* | Kislev | November/December |
| *Ṭebētu* | Tebet | December/January |
| *Šabāṭu* | Shevat | January/February |
| *Addaru* | Adar | February/March |

* Spellings are as they appear in Hunger and Pingree, 1989.

APPENDIX THREE

# TABLET AND LINE CORRESPONDENCES WITH HUNGER-PINGREE

| Hunger & Pingree's divisions | | This volume | |
|---|---|---|---|
| a. I i 1 – ii 35 | Catalogue of stars | a. | I i 1 – ii 35 |
| b. I ii 6 – iii 12 | Dates of heliacal risings | b – d | I ii 6 – iii 48 |
| c. I iii 13 – 33 | Simultaneous risings and settings | | |
| d. I iii 34 – 48 | Time intervals between the dates of heliacal risings | | |
| | | Intermediate Section I iii 49–50 | |
| e. I iv 1 – 30 | *ziqpu* stars | e. | I iv 1 – 30 |
| | | e-1 | I iv 1–9 |
| | | e-2 | I iv 10–30 |
| f. I iv 31 – II i 8 | The Path of the Moon | f. | I iv 31 – II i 8 |
| | | f-1 | I iv 31–39 |
| | | f-2 | II i 1–8 |
| g. II i 9 – 24 | First Intercalation Scheme | g. | II i 9 – 24 |
| h. II i 25 and 68–71 | Observation of heliacal risings and wind directions | h and i II i 25–71 plus Gap A, 1–7 from H & P section j | |
| | | h-i-1 | II i 25–37 |
| | | h-i-2 | II i 38–43 |
| | | h-i-3 | II i 44–67 |
| | | h-i-4 | II i 68–71 |
| | | j-1 | Gap A1–7 |
| i. II i 38 – 67 | Planetary theory | | |

(*cont.*)

| Hunger & Pingree's divisions | | | This volume |
|---|---|---|---|
| j. | II A 1 – ii 20 | Second intercalation scheme | j-2 Gap A 8 – II ii 1–7<br>j-3 II ii 18–20 |
| k. | II ii 21 – 42 | Shadow table | k. II ii 21 – 42 |
| l. | II ii 43 – iii 15 | Water clock | l. II ii 43 – iii 15 |
| m. | II iii 16 – iv 12 | Omina | m. II iii 16 – iv 12 |

# SUBJECT INDEX

# AUTHOR INDEX

# AKKADIAN AND SUMERIAN WORD INDEX

For an index of star names in MUL.APIN see Hunger & Pingree, 1989.

# MUL.APIN TEXT CITATION INDEX

NEO-BABYLONIAN ASTRONOMICAL TEXT, ETC.

86378,

REVERSE, COLUMN IV (CONTINUED).

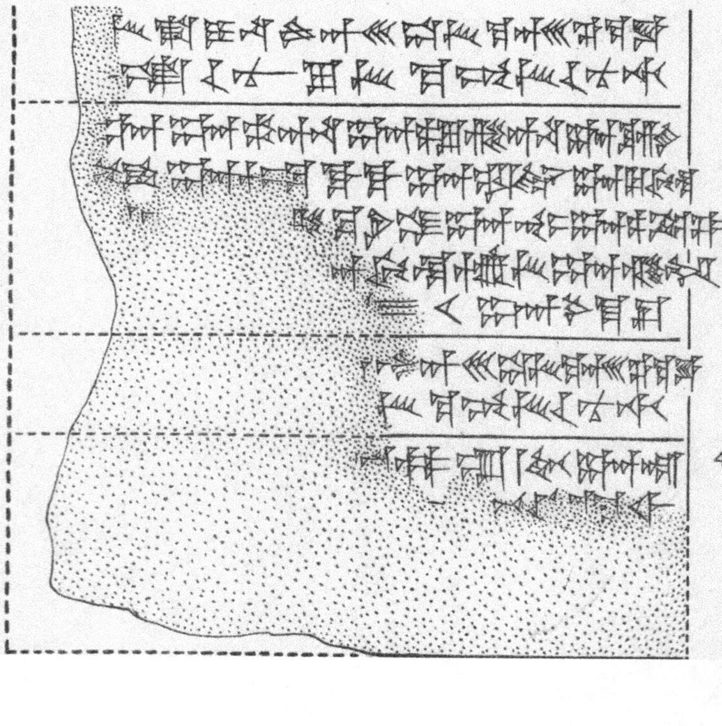

www.ingramcontent.com/pod-product-compliance
Lightning Source LLC
Chambersburg PA
CBHW071246230926
51849CB00001B/247
* 9 7 8 9 0 0 4 2 0 2 3 0 6 *